PHARMACEUTICAL PREFORMULATION

HOW TO ORDER THIS BOOK

BY PHONE: 800-233-9936 or 717-291-5609, 8AM–5PM Eastern Time

BY FAX: 717-295-4538

BY MAIL: Order Department
Technomic Publishing Company, Inc.
851 New Holland Avenue, Box 3535
Lancaster, PA 17604, U.S.A.

BY CREDIT CARD: American Express, VISA, MasterCard

BY WWW SITE: http://www.techpub.com

PHARMACEUTICAL PREFORMULATION

J. T. Carstensen, Ph.D.
Professor Emeritus
University of Wisconsin
Madison, Wisconsin

TECHNOMIC
PUBLISHING CO., INC.

LANCASTER · BASEL

Pharmaceutical Preformulation
a**TECHNOMIC**® publication

Technomic Publishing Company, Inc.
851 New Holland Avenue, Box 3535
Lancaster, Pennsylvania 17604 U.S.A.

Printed in the United States of America
10 9 8 7 6 5 4 3 2

Main entry under title:
Pharmaceutical Preformulation

A Technomic Publishing Company book
Bibliography: p.
Includes index p. 291

Library of Congress Catalog Card No. 98-86084
ISBN No. 1-56676-690-7

Dedicated, with love, to
my wife

Catherine Gene Karr

TABLE OF CONTENTS

PREFACE

This book addresses problems and solutions of formulation and preformulation with which I have concerned myself for 34 years. When I was employed in the pharmaceutical industry I worked at functions, in the 1960's, which were the precursurs of preformulation, and my early publications dealt with such matters. In the following decades advances have been made in methodology and the realm of preformulation has grown. Theory and the way in which problems are viewed have also undergone change.

The text deals with the *pharmaceutical* aspects of preformulation, not the synthetic nor the analytical aspects. It takes it vantage point at the point in time when the pharmaceutical preformulator first obtains a sample of the drug substance, and it explores the physical, chemical and technological aspects that are needed for a full exploration of the potential advantages and disadvantages of the drug substance. It is only through the understanding of underlying principles that adequate exploration can be carried out.

The "fast-track" approach to drug development advocated by many pharmaceutical companies today is an unhealthy concept. By attempting to whip scientists into meeting unrealistic deadlines, companies do themselves a disservice. Because whatever is missed at the beginning of the development of a product will come back to haunt at later times. Massive production costs on inadequate batch procedures may result in costs, and though it is granted that early arrival at the market place may result in gains that otherwise would be lost, the role of preformulation is still of great importance, because it is oftentimes possible to "repair" a process prior to NDA filing.

The book should be of interest to industrial and academic formulators, preformulators, universities, government, and patent litigators. In the latter case the point is that the manner in which the law views the process of product development and what constitutes the difference between inventiveness on one hand and routine exploration on the other can use all the reference support that can be mustered.

The author is indebted to Dr. Miriam Franchini, Dr. Madhu Pudipeddi and Dr. Mandar Dali for valuable technical assistance.

Jens Thurø Carstensen
Monona, Wisconsin,
14 March, 1998

CHAPTER 1

Introduction

Preformulation is a branch of the pharmaceutical sciences that applies to work done on a compound which is considered for development and marketing by a company. The company can be either an innovator company, i.e., a company that develops potential drug candidates and eventually markets some of them, or it can be generic, in which case a company will attempt to market an already existing drug molecule (at the time of patent expiry). Preformulation is different in these two cases, and most of this work will concentrate on the preformulation efforts by the innovator company.

1.1 THE INNOVATOR COMPANY

In an innovator company, potential drug substances are developed by either organic chemists, pharmacognocists, microbiologists or by acquisition. The methods of screening substances for pharmacological activity are beyond the scope of this book, but it is obvious that in the development of the drug substance, in the *inventive* (discovery) phase, not only is it necessary to "come up with" a compound, one must also characterize it. The first characterization is *chemical characterization*, in essence carrying out investigations that lead to a chemical structure. Tools such as NMR (nuclear magnetic resonance), MS (mass spectroscopy), HPLC (high precision liquid chromatography), as well as purely chemical techniques (functional group analysis) are used in this phase, and the work is usually carried out prior to the point when preformulation per se starts. At this point cursory physical properties are also determined (e.g. melting point, gross assessment of solubility, indications of stability to heat, alkalinity, acidity, and light).

The mode of conveying this by the discovery group to other groups (teams) that will be involved in the further development can be done in a series of ways, and here it will simply be assumed that a document is issued giving the following characteristics:

Chemical formula
Appearance
Melting point
Solubility in water and some solvents
Gross stability characteristics (hydrolysis, light, oxidation)
Physical characteristics (e.g. gross assessment of
hygroscopicity)
Toxicity

At this point, the group that has produced the substance is the discovery group. The material is not well characterized in the usual pharmaceutical sense (for instance it could be a rather coarse powder).

The work up to this point has simply been directed towards (a) discovering the compound and developing method for making the potential drug substance, and (b) showing its utility.

It is at this point in time (Year Zero) that a patent application for the drug substance is usually applied for.

At a later time (and this could be years later), a product is developed and has gone through clinical trials. Usually there is something unique about the formulation, *and it at this point in time (which could be years later) that a patent application for the product is applied for.*

1.2 PATENT CONSIDERATIONS

Generic development used to start at time zero plus seventeen and now (with the new patent rules) will probably start at time zero plus twenty. But it is obvious that the generic formula could infringe on the innovator's, and there are therefore two avenues left open for the generic manufacturer: either attempts to show that the innovator's product patent is obvious and hence should be revoked, or to develop a formula that does not infringe on the innovator's product formula.

The generic company must submit an ANDA (abbreviated new drug application), and before the FDA (Food and Drug Administration) even considers the application, it informs the innovator, who then in turn (usually) sues the generic company.

Many product patents are obvious to the pharmaceutical formulator. The problem is that patent examiners are NOT formulators, and a large body of knowledge that is second nature to the formulator is only accessible to the examiner in literature search. Any good formulator or preformulator will have determined optimum pH-values in early stages as a matter of routine and such investigations are, or should not be patentable. And if a generic company faces such a patent, the task of showing obviousness is usually the route to take.

1.3 BIOLOGICAL EQUIVALENCE AND THE PRINCIPLE OF (PATENT) EQUIVALENCE

A generic company, with its ANDA, supplies the FDA with documentation that the product they have developed is "as good" as the innovator's, and part of this documentation is what is referred to as biological (or biopharmaceutical) equivalence. The innovator's drug product and that of the generic company are administered to a group of patients, and the ensuing blood level curves of the drug in the patients should be "the same." However, it is not possible that they would be "the same," and the FDA has, therefore, established criteria for comparing blood level curves that are *equivalent*. This term causes some confusion in patent considerations.

There is a patent term that deals with the following: If a patent exists and specifies for instance the use of CMC (carboxymethyl cellulose) as a thickener, and if a generic company develops a formula using, for example, hydroxypropyl cellulose as a thickener, then the two thickeners are "equivalent" and the patent is still considered infringed upon.

The ambiguity arises, in patent cases, where the generic company shows biological equivalence, which it by law must show, and this is confused with the principle of equivalence. It is important, in patent cases, to ascertain that the judge, or the jury, understands that the two principles are different. One is a prerequisite for the ANDA, i.e. the generic company *must* establish it, and as such it has nothing to do with the principle of (patent) equivalence.

If the attainment of biological equivalence were indeed considered a part of the principle of (patent) equivalence, then a generic company would be prevented from entering the market until Year X + 20, where X is the year of the innovator's product patent issuance.

1.4 PHARMACEUTICAL PREFORMULATION

Historically, preformulation had its birth in the very early sixties. Prior to this time, many assays that were used to assess stability (e.g., those in the USP, the United States Pharmacopoeia) simply indicated the amount of drug that originally was incorporated into the dosage form. However, starting with the early sixties more and more stability-indicating assays were being introduced into pharmaceutical undertakings, and more and more instances would surface where a formulator had used his skill to make a "good" tablet of a new compound, just to find out half a year later that it was not stable. The subsequent "patch-up" work eventually led to much more work than would have been required if the job had been done "right" to start with. "Right" is a qualitative word, because the formulator, at the time, was indeed doing things in a way which, until that time, had been considered right.

But it was apparent that work had to be done prior to the point when formulation was initiated, and that was the beginning of the branch of the pharmaceutical sciences which, today, is known as preformulation.

Due to the nature of the process by which "first samples" of a new drug substance is carried out (e.g., in an organic chemistry laboratory), it is obvious that not much drug is available at the time preformulation starts. The writer, when in Industry in the sixties, spent many efforts trying to convince his superiors to supply more drug at the time when preformulation was to start.

The information that the preformulator must supply is, in part:

1. Solubility of the compound.
2. What the preferred salt forms would be.
3. Does the compound form hydrates [i.e., must it be stored in given RH (relative humidity) ranges]?
4. What is the preferable polymorphic form?

5. Is the compound hygroscopic?
6. What is the pK of the compound?
7. What is the kinetic pH-profile?
8. What would be a preferable dissolution method (medium)?
9. Compatibility with other compounds.
10. Micromeritic considerations (shape factors and crystal habits, for instance).

This list characterizes some of the difficulties in preformulation. It is obvious that, given 0.5 gram of drug substance, it would be impossible to answer all of the questions.

Secondly, some of the investigations of the topics above cannot be logically carried out unless other information is available. For instance in the selection of a dissolution method it is necessary (if correlation is to be established) that there be a collaboration with biopharmaceuticists.

This is one of the reason that many companies have adopted the team concept of product development. Ideally, then, all groups involved with any of the above functions, would "work together" in a team, and facilitate not only meaningful approaches, but also accelerate them to reduce the time from conception to marketing.

The necessity of preformulation is not only acknowledged by most companies, it is to a certain extent, a prerequisite for the IND (investigative new drug) application.

Preformulation is not confined to innovator companies. From the *generic* developer's point of view, preformulation is more a task of finding the formula that most resembles that of the innovator, and ascertaining that it, physically and stability-wise, matches (not is superior to) that of the inventor.

The other conclusion from the above list is that there are several phases to preformulation. One phase occurs early in the game for the innovator scientist, where only small amounts are available, and estimates have to be made, so that rational formulation can proceed. Because, *in the early phases.* that is the intent of preformulation, to lay some ground rule for the formulator so that a stable and efficacious product can be made for the clinic.

1.5 MOLECULAR INVESTIGATION IN LATER PHASES

In subsequent phases, at a later point in time, the rough findings obtained in early studies should be embellished upon, done "right" as they would have been had the work been a Ph.D. thesis. Somewhat shortsightedly, many companies never do the second phase.

It is shortsighted because if you called a company that had a product on the market and asked the technical staff what the solubility of the compound was, wouldn't you think that these people, who after all are the only ones in the world to have a deep knowledge of the drug substance, would be able to answer? There have been instances where the author has called such companies and got no answer and had to determine the solubility himself.

1.6 MACROSCOPIC PROPERTY DETERMINATION IN LATER PHASES

In the development of a product, after it has been determined in early clinical phases that the drug substance is a potential market candidate, thoughts of scale-up come up, eventually leading to pivotal and pilot batches.

The preformulator plays a part in this as well, because synthesis of the drug substance may affect polymorphism as well as the machinability (e.g. compression) of the substance.

A great deal can be learned by model experimentation on the drug substance itself to abort problems in later stages, so that when adequate amounts of drug substance become available several mechanical and thermodynamic determinations are in order. These include such investigations as thermal properties (in the case of polymorphism) and shape factor and particle size distribution determinations.

1.7 STATISTICS AND OPTIMIZATION OF PROTOCOLS

It is obvious that in the early phases, the utilization of a small amount of powder to yield maximum information requires statistical approaches (protocols, experimental design), and this will be touched on in the following sections and chapters as well. This, and other considerations, often require the use of the computer.

1.8 COMPUTER PROGRAMS

There are a series of examples and answers in this text using BASIC. Many good computational programs exist, but BASIC is not only the simplest, it is available at a cost that is within everybody's reach. In BASIC there are the following commands which are used in and shown whenever programs are included:

INPUT Command. The command **INPUT** X, when the program is run will print out ? and the value will have to be typed in.

Text Print. Anything which is placed between quotation marks after the **PRINT** command will print out. For example **PRINT** "X","Y" will, when the **RUN** command is executed, print

X Y

If a line is desired under this, then **PRINT** "_____________" will make the print out-put:

<u>X Y</u>

If the quotes are associated with an **INPUT** statement, then it should be followed by a semicolon.

Commas after **PRINT** separate the values by a certain distance (as shown under Text Print above).

Arithmetic Operations.

PRINT 5*10 followed by **RUN** will print out 50.

PRINT X1 will print out the value of X1 at the point in the program where the command is inserted. For instance if the program were

INPUT "X="; X1
X2 = 5*X1
PRINT X1, X2

Then running the program will first ask for the value of X1 and this is typed in:

X1 = 5

and then

5,25

will print out. If, by mistake, a symbol has not been defined, then the program and the print will assume it to be zero.

Alphabetic symbols are usually done by a letter followed by a number. For instance X1 is one variable, X2 is another, and so on up to X9.

Operational symbols are *, /, + and - and parentheses have to be placed arithmetically correctly (although this is not always necessary). Examples are

PRINT 2*3
6
PRINT 12/3
4
PRINT 2+3
5

```
PRINT 2-3
      -1
```

Arithmetical symbols. The most common symbols are *power*, ^, *exponent*, **exp**, *parenthesis, (), and natural logarithm*, **log**(), for example,

```
PRINT 2^3
      8
PRINT 2*(3+5)
      16
PRINT EXP (-1)
      0.368
LOG (10)
      2.303
```

Operational Shortcuts. The sequence: **FOR X = ...TO...STEP....** followed by **NEXT X** gives a series of values of a formula for the x-values stated. For instance

```
FOR X = 0 TO 10 STEP 1
PRINT 5*X
NEXT X
```

will produce the values 0,5,10,.....50 when the **RUN** command is given.

These short instructions should allow even the uninitiated reader to follow the programs listed in the text.

1.9 ANALYTICAL TECHNIQUES

This is not a book in analysis, but analytical techniques will be referred to, and they are listed below with a short description.

TLC - thin layer chromatography. In this technique a small drop of a solution of the material to be assayed is placed at the base of a plate containing a layer of adsorbent, usually silica or alumina. The plate is placed with the adsorbent base in a bath of the solvent, which then wicks upwards. Various compounds (e.g., a drug and its decomposition products) will migrate at different speeds. When the front of the solvent has reached a certain point, the plate is removed and dried, and the spots "developed" in various ways (depending on the compound).

There have been attempts (with some limited amount of success) to make separations of drug and decomposition products quantitative, but in general it is a qualitative technique.

High precision liquid chromatography. A column of adsorbent is placed in a tube (or bought pre-made), and the sample solution injected at one end of the tube. A solvent is then passed through the column. Various compounds (e.g. a drug and its decomposition products) will partition at different speeds onto the column. By placing a detector (e.g. a UV detector) at the exit end of the column, peaks will occur from time to time when the various compounds elute. If a standard is used, then the area under the curve of the sample can be compared with the area (and known quantity) of the standard, and the amount retained can be calculated.

The same is true for decomposition peaks if they are known. Unknown peaks (sometimes referred to as "related compounds") are sometimes "quantitated" by comparing their areas with the area under the main peak and expressing the ratio as a "percent decomposition." This is rarely correct, since the decomposition product (e.g. if a UV detector is used) rarely has the same molar absorptivity at the detection peak as does the parent compound. It is, however, a common practice.

Differential scanning calorimetry. In this technique a sample is placed in a calorimeter with a programmed temperature device, so that the temperature will increase at a given rate. The apparatus also contains a control cell with a known heat capacity. The heat flowing into the two cells is monitored and balanced to give a "baseline." If there is a thermal event in the sample cell (e.g. melting), then there will be an endotherm, which will register as a peak which falls back to baseline when the event is over. The area under the peak can be converted to calories or Joules.

Thermogravimetric analysis. In this technique the weight of a sample is monitored in the instrument that is equipped with programmed temperature increase. A peak will occur when moisture is lost, if it is, for example, lost at a given temperature (hydrates, for instance). Otherwise there will be a gradual signal. Again: areas under the peaks may be converted into weights.

Both of the thermal methods mentioned may be operated at fixed temperatures, if desired, so that the weight or heat event can be monitored kinetically, on a time scale basis.

X-Ray diffraction will be described in more detail in the text.

Microcalorimetry is a technique used more and more. A thermal activation monitor is a highly sensitive (microwatt range) instrument that is capable of monitoring heat events that are exceedingly small. By this means a variety of phenomena (e.g., adsorption, heat of solution, dissolution, crystallization) may be investigated at a high level of precision and with small samples.

CHAPTER 2

Solubility

In order to assess properties of solutions it is necessary first to be able to produce them. For this reason, it is of importance to determine the solubility of a compound in aqueous media.

2.1 AQUEOUS SOLUBILITY OF NON-IONIZED DRUGS

In the following, solubility of substances that do not ionize in solution will be treated first. The solubility of a compound should be expressed as grams (or moles) of solute per 1000 g of solvent, not per 1000 mL of solution. It is often difficult to determine solubility with small amounts of drug substance, particularly if the drug is quite soluble. However, if it is quite soluble, then many problems with formulation may not exist, other than the fact that the drug may be hygroscopic, and in such a case when hygroscopicity is determined (and this is described in a later chapter), the solubility may be deduced.

The best manner of determining solubility with small amounts of drug at hand is to use a small amount of liquid (e.g. 0.1-1 mL), and add solid until a particle no longer dissolves.

If the drug substance is fairly soluble, the drug may be regenerated by removing the water in vacuo. Most often a problematic drug substance will be poorly soluble, and in that case, after the first addition of solid which, then, does not dissolve, more water may be added. If the drug has not dissolved when a full mL has been produced, it is centrifuged, and the supernatant assayed for drug substance.

At a later stage in development of the product, when more drug substance is available, the solubilities should be determined in more conventional fashion.

It is often desirable to know the solubility at different temperatures. The most economic way of doing this is to determine the heat of solution (ΔH) calorimetrically (e.g., using a TAM - thermal activity monitor). Very small heat events may be monitored in this, allowing the heat to be measured with a minimum quantity of solid. The relation to temperature is the so-called Van't Hoff equation:

$$\ln[S] = \ln[Z] - \Delta H/RT \tag{2.1}$$

where R is the gas constant (1.99 cal/deg-mole) and T is absolute temperature (°K). The constant ln[Z] can be determined by inserting S_{25} and T = 298° in the equation, and the solubility can now be calculated for other temperatures.

Eq. 2.1 is derived from the derivative equation:

$$d[\ln S]/dt = \Delta H/RT^2 \tag{2.2}$$

where S is the solublity. It is often assumed that ΔH is independent of temperature. However, since heat capacity, c_p, is defined as:

$$(dH/dT)_p = c_p \tag{2.3}$$

it follows that, for a dissolution process:

$$(d\Delta H/dT)_p = \Delta c_p \qquad (2.4)$$

where ΔH is the heat of solution. Hence

$$\Delta H = a + (\Delta c_p)T \qquad (2.5)$$

The derivative equation now becomes:

$$d[\ln S]/dT = a/RT^2 + [(\Delta c_p)/(RT)] \qquad (2.6)$$

which integrates to

$$\ln[S] = Z - a/RT + (\Delta c_p)R\ln[T] \qquad (2.7)$$

The solubility should always be given per gram (or per 1000 g) of solvent, because this quantity is not temperature dependent.

Example 2.1

If the solubility of a compound is 2 mg/g of water at 25°C, and the heat of solution is 8,000 cal/mole, what will be the solubility at 37°C?

Answer 2.1

Equation 2.1 is programmed in the program shown (in BASIC) in Table 2.1.

TABLE 2.1 Program for Solubility

```
100 INPUT "Solubility at 25°C=";S1
110 INPUT "Heat of Solution, cal/mole="; H1
120 INPUT "Desired Temp, °C=";T1
130 X1 = LOG(S1)
140 T2 = 1/(273.15+T1)
150 T3 = 1/(298.15)
160 X2 = -(H1/1.99)*(T2-T3)
170 X3 = X1 + X2
180 X4 = EXP(X3)
200 PRINT "Temperature, °C", "Solubility, mg/g
210 PRINT T1, X4
```

When 2, 8,000 and 37 are input, then the result will be as shown in Table 2.2. The unit of the solubility will be the same as the unit in which the solubility was input.

2.2 SOLUBILITY DETERMINATIONS BY RH-RELATIONS

Most often the problem with drug substances is that they are not very soluble. However, the opposite may arise as well. In this case the following procedure may be of help, when only small amounts of drug substance are available.

When solutions are dilute, they often adhere to colligative behavior, i.e. changes in solution properties will be proportional to the mole fraction of the solute. Although drug substances may have high solubilities (e.g. 400 mg/g of water), in terms of mole fraction, X, the number is small.

TABLE 2.2 Results from Using Table 2.1

Temperature, °C	Solubility, mg/g
37	3.36974

In the cited case the mole fraction, if the molecular weight were 200, would be 2 micromoles per (55.5+0.2) micromoles of total, i.e., less than 0.04.

The vapor pressure, P, over the solution would be given by:

$$P = P_o(1-X) \tag{2.8}$$

where P_o is pure water's vapor pressure. Since water activity, a, is defined as:

$$a = P/P_o \tag{2.9}$$

and relative humidity, RH, is

$$RH = 100a = 100(1-X) \tag{2.10}$$

it follows that the RH over a solution is a function of the concentration in a linear fashion as shown in Figure 2.1.

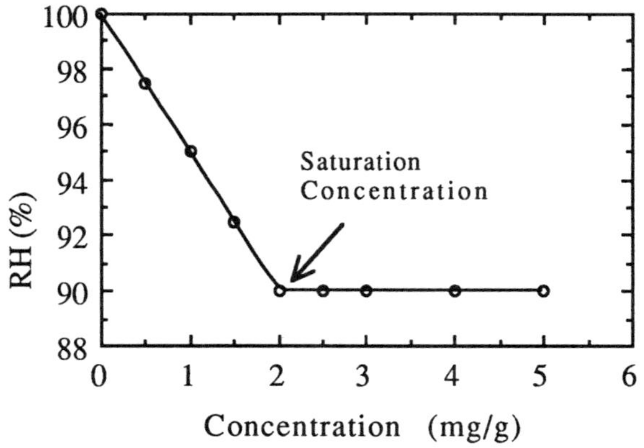

FIGURE 2.1 Ideal vapor pressure behavior of a solute in aqueous solution.

If a substance is placed in a moist atmosphere it will pick up moisture assuming that the water vapor pressure, P, is larger than that of a saturated solution. This allows calculation of the RH value over a saturated solution. This has, for instance, been carried out by VanCampen et al. 1980).

These data are plotted in Figure 2.2. It is obvious by visual observation that the moisture uptake rate starts increasing at 70% RH. It will be shown later that this RH-value corresponds to that over a saturated solution. From this information it is, therefore, possible to calculate the mole fraction of the solid (and hence the mg per g of water) at saturation.

Example 2.2

If, in the data in Table 2.1 and Fig. 2.2 the molecular weight of the compound were 400 mg/g of water, what would an estimate of its solubility be?

TABLE 2.3 Moisture Uptake Rates Versus Relative Humidity

Relative Humidity (%)	Moisture Uptake Rate (mg/g/day)
30	0.01
40	0.02
50	0.025
60	0.027
70	0.03
75	10
80	20
85	30

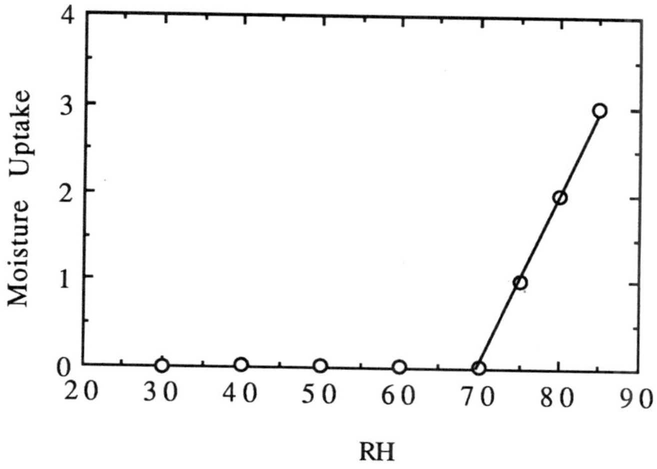

FIGURE 2.2 Moisture uptake rates from Table 2.2.

Answer 2.2

Using Equation 2.10 it follows that at saturation:

$$RH_s = 100(1-X_s) \tag{2.11}$$

where the subscript s denotes "at saturation." Hence:

$$70 = 100(1-X_s) \tag{2.12}$$

so that

$$X_s = 0.3 \tag{2.13}$$

If there are N g of drug substance per g of water, then:

$$0.3 = (N_s/400)/[(1/18)+(N_s/400)] \tag{2.14}$$

or:

$$0.3/18 + 0.3[N_S/400] = N_S/400$$

or

$$N_S = \text{ca } 10 \text{ g/g of water} \quad (2.15)$$

It is obvious that such a high solubility could not be achieved by conventional means. For example, such very soluble compounds (as, for instance, ranitidine hydrochloride) in concentrated solutions are very gel-like. Even with large amounts of drug substance available, filtration is often impossible, and centrifugation may also be problematic.

The above method assumes ideality. If the critical relative humidity is less than 70%, then this requirement is often not met, and the method gives, at best, an estimate.

It should be noted that the "molecular weight" used should be divided by two if the substance is fully dissociated into two ions (as for instance an amine hydrochloride), or by three (or Q), if three (or Q) ions are the result of the dissociation.

An instrument called a VTI is often used to determine moisture adsorption isotherms. Only very small amounts of drug are necessary for such experiments. A modified TAM (Pudipeddi et al., 1995, 1996) may also be used.

2.3 SOLUBILITY IN MIXED SOLVENTS

In the above case, the experiment may be repeated, but in place of water, a suitable, mixed solvent (ethanol, propylene glycol) may be used in one or two concentrations (25%, 50%). If the solubility in the pure solvent is also known, then, denoting the solubilities S_{25}, S_{50}, and S_{100} the solubility, S, over the whole range may be estimated by assuming a second-order polynomial relation-ship:

$$S = a + bX + cX^2 \quad (2.16)$$

Example 2.3

The solubility of a poorly soluble compound is 0.01 mg/mL in water and is tested in aqueous solutions of PG of various concentrations. The results are shown in Table 2.4. What would it be in 35% propylene glycol?

TABLE 2.4 Solubility of a Drug Substance in Aqueous Propylene Glycol Solutions

Weight Fraction PG	Solubility
0	0.05
0.25	1.1
0.5	1.9
1	1.5

The data in Table 2.4 and the solubility for two other drug substances are shown in Figure 2.3.

It is interesting to note (Paruta and Irani., 1964) that if the dielectric constant of the liquid is known at the maximum solubility (Cavé et al., 1979), then any other liquid pair with a similar dielectric constant will provide maximum solubility in that system.

These types of graphs may also be accomplished by, in place of the dielectric constant, using Hildebrand solubility parameters. This will be discussed in further detail in a following section.

Answer 2.4

Values for a, b and c may be obtained by use of a graphics program (e.g. Cricket III), as shown in the lower curve in Figure 2.3. The curve need not have a maximum, but when it does, *it has the advantage that other solvents with the same dielectric constant as that at the maximum will give maximum solubility in that system.*

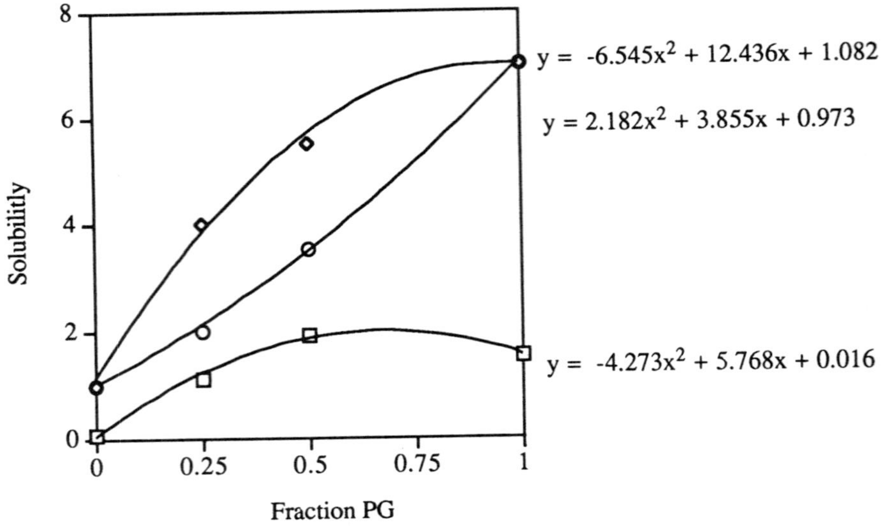

FIGURE 2.3 Solubility of a drug in aqueous propylene glycol. The lower curve (with the symbols) refers to Example 2.3.

2.4 SOLUBILITY OF DRUG SUBSTANCES WHICH IONIZE IN SOLUTION

Strong acids and bases and their salts dissociate fairly fully in aqueous solution. Weak acids (HA) dissociate partially, as shall be discussed in Chapter 3. The solubility of such situations will be discussed in that chapter, and it suffices here to say that a species (e.g. the acid anion, A^-) may precipitate in the presence of a counter ion (B^+), i.e.:

$$A^- + B^+ \leftrightarrow [AB]_{precipitate} \tag{2.17}$$

The solubility is governed by a solubility product, K_{AB}, given by:

$$K_{AB} = [A^-][B^+] \quad (2.18)$$

In case of a hydrochloride salt of an amine, for instance, increasing the chloride ion in solution may decrease the solubility of the amine ion.

2.5 IONIC STRENGTH

A strong electrolyte, e.g. NaCl, is fully dissociated in water, i.e.

$$AB \rightarrow A^+ + B^- \quad (2.19)$$

If the normality of AB is N, then the normality of both A^+ and B^- is N.
In general we may write an electrolyte as:

$$A_mB_n \rightarrow mA^{n+} + nB^{m-} \quad (2.20)$$

The effect of an electrolyte on various properties of a solution is its ionic strength; this is denoted μ and is defined as:

$$\mu = 0.5\ \Sigma[m(n_+{}^2) + n(m_-)^2] \quad (2.21)$$

<u>Example 2.4</u>

What is the ionic strength of a 1.5 normal solution of calcium chloride?

<u>Answer 2.4</u>

$$CaCl_2 = Ca^{++} + 2Cl^- \quad (2.22)$$

The molarity of the calcium ion is 1.5, that of the chloride ion is 2 x 1.5 = 3, so that the ionic strength is

$$\mu = 0.5\ \Sigma[1.5x4 + 3x1] = 4,5 \qquad (2.23)$$

since the charge on the calcium ion is 2 and the charge on the chloride ion is (minus) one.

For instance, K_{AB} is a function of ionic strength.

2.6 SOLUBILITY DETERMINATIONS WHEN ADEQUATE SUPPLIES OF DRUG SUBSTANCE ARE AVAILABLE

The proper method for determining solubility of a compound is to place an excess of solid in contact with the liquid in question. This could be stored in a tight bottle, and this bottle then rotated in a constant temperature water bath. Samples of the supernatant would then be obtained every hour for 8 hours, then every 24 hours for 4 days. The samples would be either filtered (in cases where there is no problem with adsorption on the filter paper) or centrifuged.

The supernatant is then assayed for content of drug substance.

A typical set of data are shown in Table 2.6 and Figure 2.4. The solubility may be read directly off the graph or from the Table (10.5 mg/mL), but if the data are collected at even intervals, then the so-called Nogami method may be used (Nogami, 1966). If the concentration column is "moved one down," and the last column plotted versus the second column, then a straight line should ensue, and the solubility is the point where this line intersects with the line y = x. The data plotted in this fashion are shown in Figure 2.5.

It is seen that the line cuts the line y = x when $1.3387 = 0.1275x$, i.e., when $x = y = 1.3387/0.1275 = 10.5$ mg/mL.

TABLE 2.5 Results from a Solubility Study

Time (hr)	C at t	C at t-q
0.000	0.000	
1.000	1.187	0.000
2.000	2.240	1.187
3.000	3.174	2.240
4.000	4.003	3.174
5.000	4.737	4.003
6.000	5.389	4.737
7.000	5.967	5.389
8.000	6.480	5.967

It should be pointed out that when a metastable polymorph is at hand, then the plots may not come out as shown in Figure 2.4, but may go through a maximum. This will bediscussed further at a later point in the chapter dealing with polymorphism.

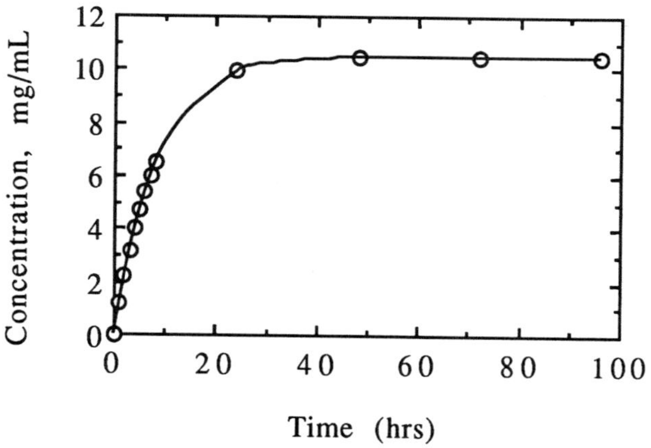

FIGURE 2.4 Data from Table 2.5. The data from 24, 48, 72 and 96 hours are not shown in the table.

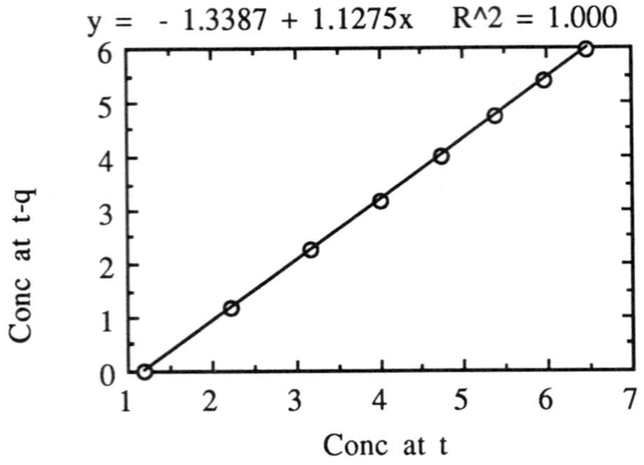

FIGURE 2.5 Data from Table 2.5 where the last column is plotted versus the next-to-last column.

2.7 SOLUBILITY OF POORLY STABLE COMPOUNDS

When a compound is poorly stable in aqueous solution, then the Nogami method may be particularly useful, because the longer the solubility experiment goes on, the more drug substance will degrade. Then it is worthwhile selecting a smaller value for q, for instance, 20 minutes.

The method of approaching this is as follows. The rate with which the compound goes into solution in V mL of liquid is given by:

$$dM/dt = -VdC/dt = -kAS \tag{2.24}$$

or

$$dC/dt = kAS/V \tag{2.25}$$

where M is mass not dissolved at time t, A is the surface area, S is the solubility and k is the intrinsic dissolution rate constant.

The rate with which the drug decomposes is given by:

$$dC/dt = -k_1 C \quad (2.26)$$

where first-order kinetics are assumed, and where k_1 is the decomposition rate constant.

The concentration profile is therefore governed by the equation:

$$dC/dt = (kAS/V) - k_1 C = k_1(\beta - C) \quad (2.27)$$

where

$$\beta = (kAS/k_1 V) \quad (2.28)$$

This may be rewritten:

$$dC/(\beta - C) = -k_1 dt \quad (2.29)$$

which integrates to

$$\ln[\beta - C] = -k_1 t + \ln[\beta - C_o] \quad (2.30)$$

The value of β may be found by iteration, and this then gives both the value of S and k_1 (provided the surface area is known).

2.8 EFFECT OF SURFACTANTS

Surfactants, such as the polysorbates, will solubilize drug compounds when the surfactant is present in excess of its critical micelle concentration (cmc). Beyond this, the solubility will increase linearly with surfactant concentration.

Surfactants also aid in the wetting down of a solvent, and they are often used in dissolution tests for poorly soluble drugs.

There will be more description of this in the next chapter.

2.9 MIXED SOLVENT SYSTEMS AS A FORMULATION AID IN SOLUBILIZATION

When a drug substance is poorly water-soluble it often falls upon the preformulator to design solution systems which achieve an adequate concentration for a solution dosage form.

There are three prime approaches to this: (a) solubilization by means of co-solvents, (b) complexation and (c) micellar solubilization. The latter will be treated in a later chapter, and the approach of mixed solvents will be discussed here.

The topic has, to some degree, been discussed in section 2.3, but some added comments are in order here. Co-solvents used are
ethanol, glycerin, propylene glycol, and polyoxyethylene glycols.

Changing the aqueous phase drastically (e.g. adding large percentages of sugar) is also an approach, because, as mentioned earlier, one of the factors that is of great importance is the dielectric constant. If graphs such as Figure 2.3 are obtained, then the dielectric constant of the composition that gives the maximum solubility may be obtained, and then other solvent systems may be formulated to this dielectric constant. This obviates a lot of solubility determinations.

If more than one co-solvent is used, then ternary diagrams may be employed to visualize where maximum solubility occurs.

The distance PA is the amount of water, the distance WB is the percentage of Ethanol and the distance EC is the amount of propylene glycol. In the ternary diagram in Figure 2.6 the lines

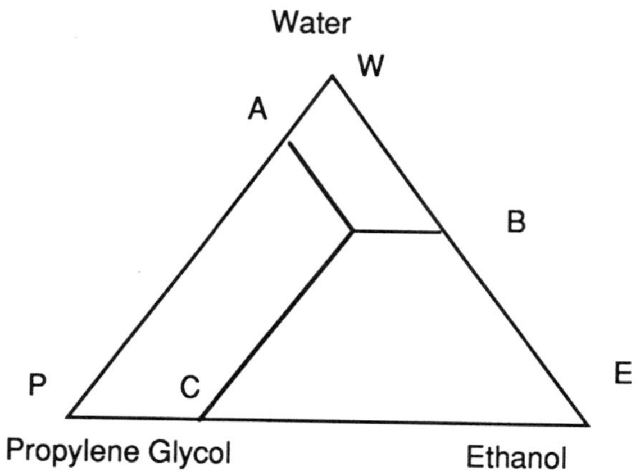

FIGURE 2.6 Ternary diagram.

are drawn parallel to the sides of the triangle. They could, as well, be drawn perpendicularly to the axis. Any point within the triangle then defines a given composition.

If, in an extended study, solubilities are determined for say 100 compositions, then the compositions for which the solubility was the same (1 mg/mL, 2 mg/mL, etc.) might be connected and one of two types of diagrams would result as shown in Figure 2.7. The one diagram in the left of the figure results in a maximum solubility, whereas the other diagram simply implies that the more of one of the cosolvents is added, the higher the solubility.

The diagrams are pictorial and condense a large amount of material into one figure. However, large amounts of experiments may not be done in the early stage of preformulation, so in these cases it is better to express the solubility mathematically.

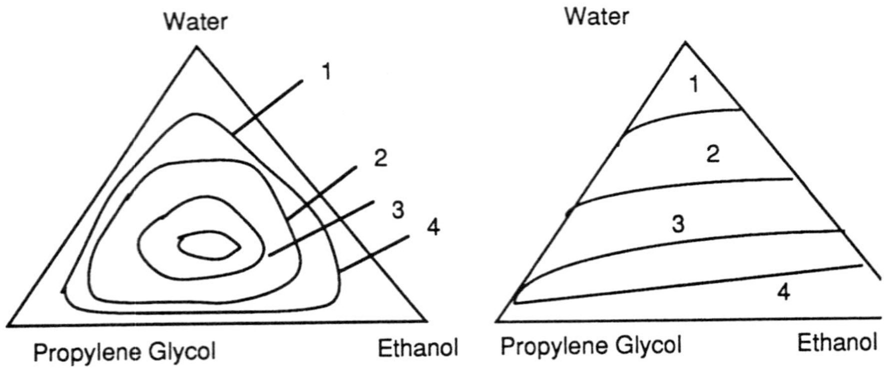

FIGURE 2.7 The two types of solubility ternary diagrams.

2.10 COMPLEXATION AS AN AID IN SOLUBILIZATION

Drug substances may complex with complexing agents. An example e.g. is ascorbic acid/nicacinamide. In general, one of the two components of the system, e.g. the drug A, is called the substrate and the other, B, is called the ligand,

$$A + B <-> AB \quad (2.31)$$

The equilibrium constant (the "stability" constant) is given by:

$$K = [AB]/[A][B] \quad (2.32)$$

There are several manners of establishing (a) that the complexation is 1:1 as shown in Eq. 2.31 and (b) what the equilibrium constant is. Two of the methods will be dealt with here, and the third, the kinetic method, will be touched on in the chapter on solution kinetics.

In the solubility method, the solubility, S_{obs}, is determined at different levels of ligand, [B]. The solubility in the absence of ligand, S, is known, or has previously been determined. A certain amount will complex, so that the solubility will increase (or at times decrease) over and beyond the value of S. In Eq. 2.32, the actual concentration of B will be the added amount [B] minus the complexed amount, [AB], so that Eq. 2.32 may be rewritten:

$$K = [AB]/\{S([B]-[AB])\} \tag{2.33}$$

or

$$[AB] = KS[B]/\{1+KS\} \tag{2.34}$$

The observed concentration

$$S_{obs} = S + [AB] = S + KS[B]/\{1+KS\} \tag{2.35}$$

or

$$S_{obs}-S = KS[B]/\{1+KS\} \tag{2.36}$$

If the observed increase in solubility is plotted versus the added concentration of ligand, the plot will be linear, with a slope of $KS/\{1+KS\}$ so that K may be determined.

Most often, AB has a different ultraviolet (UV) spectrum from that of a weigthed sum of A and B spectra, and K may be determined from this as well (Connors and Mollica, 1964).

The most common ligands are: cyclodextrins, caffeine, urea and polyoxyethylene glycols. However, other complexing agents have been used successfully, e.g. N-methyl glucamide. Often complexed species have higher biological absorption rates than the native compound.

2.11 PARTITION COEFFICIENTS

When two immiscible phases are placed in contact with one another, e.g. one being octanol and the other being water, and if then a drug substance is added to the system, then part of the drug will dissolve in the octanol phase and part in the water phase. The concentrations in the respective phases are C_{or} and C_{aq}. The ratio between the two is denoted the partition coefficient Π, i.e.

$$\Pi = C_{or}/C_{aq} \tag{2.37}$$

Octanol/water partition coefficients are important, since absorption in the body is frequently a function of Π. It should be determined at several concentrations to assure that it is indeed a constant. If there is molecular aggregation (e.g. dimerization) of the drug substance in one of the phases, then one of the concentrations would have to be raised to a power (n) in order that the ratio remain constant.

In companies where many compounds of a similar structure are tested in preformulation it is often possible to establish Hammett-type relations, which will predict the partition coefficient as a function of the "new" substituent.

2.12 DIELECTRIC CONSTANT AND SOLUBILITY PARAMETERS

Solubility in mixed solvents has been touched on in the previous writing. If several different solvent systems are being investigated, then it often happens (Lordi et al., 1964), that there will be a maximum in solubility at a given dielectric constant, regardless of which solvent pair has been used (Figure 2.8).

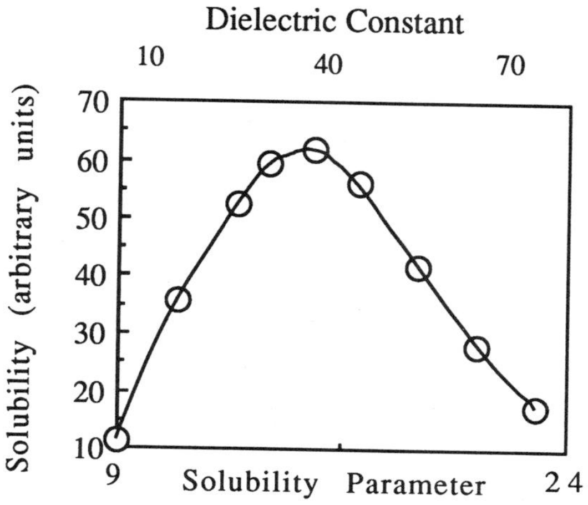

FIGURE 2.8 Example of the solubility of a compound as a function of dielectric constant and of Hildebrand solubility parameter. Graph constructed from data published by Lordi et al., 1964.

Rather than using dielectric constant as a measure, the Hildebrand solubility, δ, parameter is, at times, used. This is defined as (Shinoda, 1971):

$$\delta = \{(\Delta H - RT)/V\}^{1/2} \tag{2.38}$$

where ΔH is the heat of vaporization, V is the molar volume, R is the gas constant, and T is absolute temperature. If ΔH is not known at the temperature in question, then it may be obtained from the Clausius-Clapeyron Equation:

$$\{d\ln P/dT\}_{sat} = \Delta H/(RT^2) \tag{2.39}$$

If the surface tension, γ, is known as a function of temperature (or at the temperature in question), then it may be estimated from (Beerbower, 1971):

$$\delta = 3.741[\gamma/V^{1/3}]^{0.5} \quad (2.40)$$

It is noted that the solubility parameter for water is 23.8.

2.13 SOLUBILITY AND pH

The solubility of acids and bases is, of course, a function of pH. For a monobasic acid with a given pK the observed solubility, S, will be the sum of ionized and non-ionized parts (HA and A^-).

In the following it will be assumed that a pH is obtained by adding sufficient MOH (where M is a monovalent metal ion) to the acid to form a saturated solution.

The Henderson-Hasselbach equation states that:

$$pH = pK_a + \log_{10}\{C_{A^-}/C_{HA}\} \quad (2.41)$$

or

$$C_{HA} + C_{A^-} = C_{HA}\{1+10^{pH-pKa}\} \quad (2.42)$$

If the solubility of HA, S_{HA}, is less than that predicted by the solubility product of A^-, i.e.

$$K_1 = [A^-\}[M^+] \quad (2.43)$$

Then the observed solubility, S, of the compound at the pH in question is

$$S = S_{HA}\{1+10^{pH-pKa}\} \quad (2.44)$$

The solubility curve would, in such a case, look like a titration curve.

If, however the acid is dibasic, then three species are in question, viz. H_2A, HA^- and $A^=$. It would have two pK's, i.e., pK_1 for the pair H_2A/HA^- and pK_2 for $HA^-/A^=$. At lower pH, $A^=$ is negligible, and the shape of the curve would be FG in Figure 2.9 dictated by:

$$S = S_{HA}\{1+10^{pH-pK1}\} \qquad (2.45)$$

At higher pH the species HA would be negligible, and in this case the Henderson-Hasselbach equation would dictate that

$$C_{HA^-} = C_{A^=}\{10^{pK_2-pH}\} \qquad (2.46)$$

If the solubility of $A^=$ is smaller than that of HA^-, then

$$S_{obs} = S_{A^=} + C_{HA^-} = S_{A^=}\{10^{pK_2-pH}\} \qquad (2.47)$$

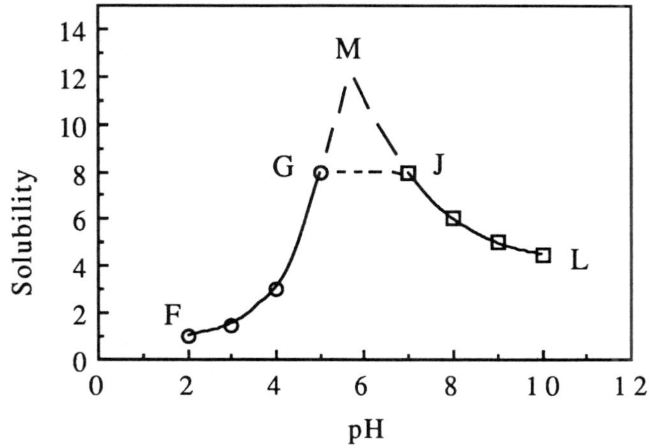

FIGURE 2.9 Profiles that may occur if M_2A is less soluble than MHA, which in turn is more soluble that H_2A.

A trace of such a curve is shown in JL in Figure 2.9.

What happens in the intermediate region is a bit more complicated, since here all three species are present, and the shapes may be as demonstrated in Figure 2.9 by either shape GMJ or simply GJ. With tribasic acids it is often the shape FGJL that prevails.

The area between G and J is often poorly duplicable.

There are other reasons than the ones stated above for profiles as shown. It is often the case that double salts occur, and these would occur in regions where, for example, HA^- and $A^=$ are present in favorable concentrations.

Finally, incongruent melting should always be taken into account (Carstensen, 1990; Carstensen and Ertell, 1989). Here the solubility is actually affected by the amount of solid material present.

2.14 CURVED VAN'T HOFF PLOTS

At times the Van't Hoff solubility plots are curved. In such cases Grant et al. (1984) have suggested that at times the reason for this is that ΔH of solution is temperature dependent, i.e., that

$$\Delta H = a + bT \qquad (2.48)$$

where a and b are constants. Denoting by s_2 the saturation concentration in a suitable concentration scale, and inserting this in the equation:

$$d\ln s_2/dT = -\Delta H/(RT^2) \qquad (2.49)$$

It is noted that

$$\Delta C_P = d\Delta H/dT \qquad (2.50)$$

where ΔC_P is the difference between heat capacities of the drug in the solid state and in saturated solution. Integrating Eq. 2.49 and employing Eq. 2.50 leads to:

$$\ln[s_2] = c - \{(a/R)/T\} + \{(b/R)\ln T\} \quad (2.51)$$

where a, b and c are constants.

Grant et al. (1984) suggest treating the equation by multiple regression, but this is statistically unsound for several reasons: (a) the dependent variable is in logarithmic transformation, and hence not normally distributed, (b) the two independent variables (1/T and lnT) are not independent, and (c) they are both transformed, and in particular the reciprocal (1/T) is not well amenable to least squares fitting.

The treatment can lead to errors, and it is better to use non-linear regression, i.e., Eq. 2.51 in the form:

$$s_2 = \{e^c e^{(-a/RT)}[T^{(b/R}] \quad (2.52)$$

Even Eq. 2.52 can be misleading if good first estimates of a, b and c are not on hand. It is noted that ΔH is not synonimous with calorimetrically determined heats of solution, since these are usually done at high dilution. Pudepeddi et al. (1996) have suggested the heat of crystallization (which can be determined calorimetrically) to be a good first estimate for ΔH.

An example of a set of data is shown in Table 2.6.

Table 2.6 Solubility Data

Temperature, °K	Mole Fraction, X	Multiple Regression Fit	Non-Linear Regression Fit
278.15	0.009275	0.0077	0.0029
288.15	0.013054	0.0144	0.0077
298.15	0.021321	0.0265	0.0192
308.15	0.038864	0.0478	0.0454
318.15	0.104152	0.0850	0.1022

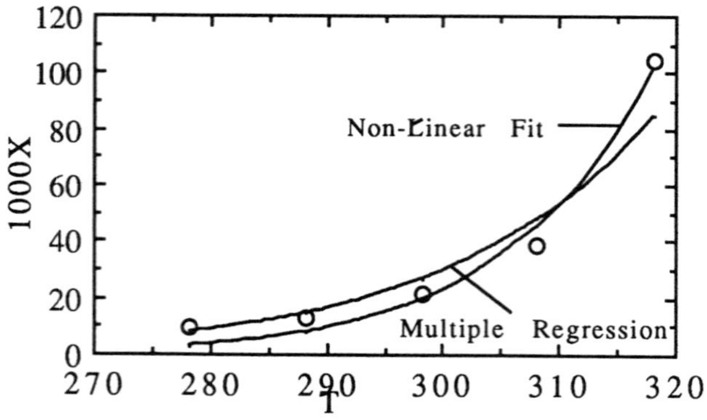

FIGURE 2.10 Data from Table 2.6 plotted by multiple regression and by non-linear regression.

Data from Table 2.6 are plotted by multiple regression and by non-linear regression in Figure 2.10. The sums of the residuals are 24.1 for the non-linear regression fit (Sigma plot program), and 95.6 for the multiple regression fit, so that from a statistical point of view the former is better. The problem is that the values for a/R and b/R (and hence the heat of solution and the change in heat capacity) are quite different for the two treatments, as shown in Table 2.7.

It should be realized that curvature is not necessarily due to temperature dependence of the heat of solution, but may also

Table 2.7 Parameters in Curve Fitting of X

	Multiple Regression	Non-Linear Regression
a/R	950	6,681
b/R	21.06	4.08
c	-126.8	0.008224

be due to the temperature dependence of the activity coefficients. The equations where X is used without an activity coefficient term are assumed to be ideal.

2.15 PREFORMULATION TASKS IN THE AREA OF SOLUBILITY

When only small amounts of material are available then the solubility should be determined with the most material saving method. For a poorly soluble compound, the solubility may be obtained by the conventional method shown in Section 2.6, i.e. by adding an excess amount of drug to a small amount of water.

Plotting according to the Nogami method (Figure 2.5) will give the solubility and is advocated, since it circumvents the effect of possible decomposition of the drug substance.

When only small amounts of material are available for a compound with high solubility, this is best assessed by plotting the moisture uptake rates in a vacuum electrobalance at different RH values and finding the relative humidity above a saturated solution in the manner shown in Figure 2.2.

REFERENCES

Beerbower, A., (1971), *J. Colloid Interface Sci.,* **35**:46.

Carstensen, J.T., (1977), *Pharmaceutics of Solids and Solid Dosage Forms*, Wiley/Interscience, New York.

Carstensen, J.T., Aron, E.G., Spera, D.C. and Vance, J.J., (1966), *J.Pharm.Sci.*, **55**: 561.

Carstensen, J.T. and Ertell, C., (1989), *Proceedings of the 5th APGI Conference, Paris, France,* May 30-June 2, 1989 Volume V:307.

Carstensen, J.T. (1990), *Drug Stability,* Marcel Dekker, New York, N.Y., pp. 194-195.

Cavé, G., Puisieux, F. and Carstensen, J.T., (1979), *J. Pharm. Sci.*, **68**: 424.

Connors, K.A. and Mollica, J.A., (1964), *J. Pharm. Sci.*, **53**:772.

Grant, D.J.W., Mehdizadeh, M., Chow, A.H.-L., and Fairbrother, J.E., (1984), *Int. J. Pharmaceutics,* **18:**25

Lordi, N., Sciarrone, B., Ambrosio, T, and Paruta, N., (1964), *J. Pharm. Sci.,* **53**: 463.

Maron, C. and Prutton, S.H., (1965), *Principles of Physical Chemistry*, 4th Ed., MacMillan, London, p. 445.

Nogami, H., Nagai, T. and Chida, H., (1966), *Chem. Pharm. Bull.*, **14:**152.

Paruta, A.N. and Irani, S.A., (1964), *J. Pharm.S ci.*, **54:**1334.

Pudipeddi, M., Dudu, S., Sokoloski, T. and Carstensen, J.T., (1995), *J. Pharm. Sci.*, **84:** 1236.

Pudipeddi, M., Dudu, S., Sokoloski, T. and Carstensen, J.T., (1996), *J. Pharm. Sci.*, **85:** 381.

Shinoda, K., (1971), *Principles of Solution and Solubility,* Marcel Dekker, New York, p.64.

VanCampen, L., Zografi, G. and Carstensen, J.T., (1980), *Int. J. Pharm.*, **5**:1.

CHAPTER 3

pH and pK

3.1 DEFINITION OF pH

pH is defined as the negative of the logarithm of the hydronium ion activity, but in this text we shall simply substitute concentrations for activities, since they are almost equal in dilute solution. Hence, pH is (Brønsted, 1943a):

$$pH = -\log_{10}[H_3O^+] \tag{3.1}$$

or

$$[H_3O^+] = 10^{-pH} \tag{3.2}$$

Some authors employ activity rather than concentration in the definition of pH. The ion $[H_3O^+]$ is usually referred to as the hydronium ion. It should be noted that it is 10-logarithms that are used. It is conventional, when strictly aqueous solutions are considered, to simplify this by simply referring to $[H^+]$, in which case the definition in Eq. 3.1 becomes:

$$pH = -\log_{10}[H^+] \qquad (3.3)$$

and

$$[H^+] = 10^{-pH} \qquad (3.4)$$

This convention will be followed in most of this text, except when special circumstances dictate otherwise. It should be pointed out, however, that if a mixed solvent medium (e.g. 20% propylene glycol) were used, then the species present (assuming solvation of H^+ by one solvent molecule) would be:

$$H_3O^+ \text{ and } C_3H_7(OH)_2^+ \qquad (3.5)$$

and since it is not known how much of each is ionized, the measurement of a "pH" with a pH-meter, for instance, does not give the correct value of $-\log[H_3O^+]$. Some adjustment towards the true value may be obtained by standardizing the electrode in a standard solution containing 20% propylene glycol.

If the compound is a weak acid, then it dissociates in aqueous solution as shown below:

$$HA \leftrightarrow H^+ + A^- \qquad (3.6)$$

and is associated with an equilibrium constant, $K_{(c)}$,[1] which is given by:

$$K_{(c)} = [H^+][A^-]/[HA] \qquad (3.7)$$

---

[1] Concentrations, rather than activities, are used in this text.

where $K_{(c)}$ is the equilibrium constant at a given concentration level when concentrations, rather than activities, are used. Characteristically, for instance, for acetic acid

$$K_a = 10^{-4.75} \tag{3.8}$$

In this case the terminology K_a has been used and implies that $K_{(c)}$-values have been determined at different concentrations and extrapolated to zero concentration (where activity = concentration). Rather than work with small numbers, the term pK_a is used, and it is:

$$pK_a = -\log_{10}[K_a] \tag{3.9}$$

so the pK_a for acetic acid would be 4.75.

If an acid is dissolved in water in a concentration of C, and if it dissociates to an extent of α, then the remaining concentration of HA will be $C(1-\alpha)$ and the concentration of both H^+ and A^- will be $C\alpha$. When this is inserted in Eq. 3.7, Eq. 3.10 results:

$$[C\alpha]^2/C(1-\alpha) = K_{(c)} \tag{3.10}$$

When α and C are small, then this becomes:

$$[C\alpha] = [H^+] \approx [CK_a]^{1/2} \tag{3.11}$$

One often talks about the "natural" pH of an acid, and this refers to the pH of an aqueous solution of the drug substance. But this pH (as seen in Eq. 2.18) is a function of concentration. If the concentration is equal to the K_a, then the hydrogen ion concentration is equal to the K_a.

3.2 OBTAINING pK_a WITH SMALL AMOUNTS OF DRUG

The conventional method for determining pH is by means of titration. The drug substance is dissolved in a suitable amount of water, and the electrode(s) of a pH-meter is inserted in the solution. By means of a burette, measured amounts of alkali of known normality are added, and the pH is recorded as a function of the volume of titrant. For an acid, the equilibrium is given by:

$$HA + H_2O <-> H_3O^+ + A^- \quad (3.12)$$

It is noted that from the equilibrium equation:

$$K_{(c)} = [H_3O^+][A^-]/[HA] \quad (3.13)$$

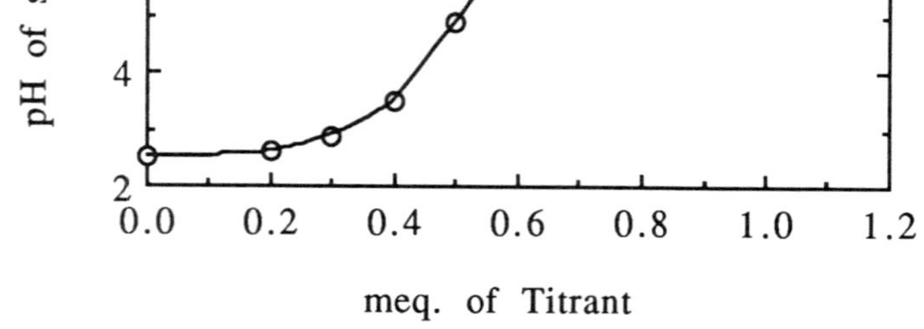

FIGURE 3.1 Meq. of base added per meq. of acid.

If activities are substituted for the concentrations, then the constant is the thermodynamic equilibrium constant, K_a. In dilute solution $K_{(c)} = K_a$, so that

$$pH = pK_c + \log\{[A^-]/[HA]\} \qquad (3.14)$$

At half-neutralization $[A^-] = [HA]$ so at this point the pH is equal to the $pK_{(c)}$ or if the solution is dilute, the pK_a. In Figure 3.1 this point corresponds to a pH of about 6, so that the $pK_a = 6$, provided the solution is very dilute.

3.3 APPROXIMATE RELATION BETWEEN pK_a AND $pK_{(c)}$

For 1:1 acids (e.g. acetic acid) the correlation between the K_a and $K_{(c)}$-values at low concentrations are often given by (Brønsted, 1943b):

$$pK_{(c)} = pK_a - z_B\sqrt{\mu} \qquad (3.15)$$

where z_B is the charge of the conjugate base. If z_A is +1 (i.e. $z_B = 0$ as in the case of amine hydrochlorides), then $pK_a = pK_{(c)}$. At higher concentrations, the correlation is

$$pK_a = pK_{(c)} - [z_B\sqrt{\mu}/(1+\sqrt{\mu})] \qquad (3.16)$$

For values of acids of the type H_hA_a the second term has to be multiplied by ha.

pK_a- values are tabulated in common handbooks, and $K_{(c)}$-values may be calculated from them by knowing the concentrations and the ionic strength of the system and applying formula 3.15 or 3.16.

3.4 SPECTROPHOTOMETRIC DETERMINATION OF pK

In Beer's law, which is used analytically to determine drug concentrations, for instance, the optical density (I/I_o) of light passing through a solution of drug in a solvent, should be proportional to concentration:

$$\log(I/I_o) = -\varepsilon a C \tag{3.17}$$

where ε is the molar extinction coefficient, a is the path length and C is the concentration.

In the case of a weak electrolyte, for instance an acid HA, it often happens that the anion (A) resonates, i.e. has a spectrum in the UV-region, whereas the undissociated acid does not. If the molar absorption coefficients are denoted ε_{HA} and ε_A, at a given pH, the absorbance, ε, will be the sum of the absorbances of HA and A, i.e. if there is a fraction x of the latter and (1-x) of the former:

$$\varepsilon = x\varepsilon_A + (1-x)\varepsilon_{HA} \tag{3.18}$$

At a given wavelength, the only unknown is x, which may then be calculated.

The determinations are carried out in a buffer system, at different pH values. x may therefore be calculated for different pH values, and by using the Henderson-Hasselbach equation (Eq. 3.14) it is possible to calculate the value of pK.

This should be done at two or three buffer concentrations, and a plot of $pK_{(c)}$ may then be plotted versus $[\sqrt{\mu}/(1+\sqrt{\mu})]$ as advocated in Eq. 3.16 to give the value of pK_a.

3.5 SOLUBILITY OF A WEAK ELECTROLYTE AS A FUNCTION OF pH

If the solubility of the anionic form of a weak electrolyte, A, is denoted S_A, and that of the undissociated acid, HA, is denoted S_{HA}, then, at any given pH, the observed solubility, S_{obs}, will be the sum of the solubility of HA and the concentration of A^-, i.e.

$$S_{obs} = S_{HA} + [A^-] \tag{3.19}$$

This assumes that $[A^-]$ does not supersede the solubility product with the cation which is present in the system.

Eq. 3.14 may be rewritten:

$$\log[A^-] = pH - pK_a + \log[HA] \tag{3.20}$$

The antilogarithmic form of Eq. 3.20 is:

$$[A\text{-}] = [HA]10^{(pH-pKa)} = S_{HA}10^{(pH-pKa)} \tag{3.21}$$

Inserting this into Eq. 3.19 gives:

$$S_{obs} = S_{HA}\{1+10^{(pH-pKa)}\} \tag{3.22}$$

An exactly similar argument may be carried out for a base, e.g., an amine hydrochloride.

3.6 PARTITION COEFFICIENTS OF WEAK ELECTROLYTES

If a weak electrolyte, e.g. an acid HA, is allowed to partition between e.g., octanol and water, then (to a first approximation) only HA will partition into the organic layer.

The observed concentration in this layer hence is

$$C_{or} = C^{or}{}_{HA} \qquad (3.23)$$

whereas in the aqueous layer (in a manner similar to the derivation of Eq. 3.21) it is:

$$C_{obs} = C^{aq}{}_{HA}\{1+10^{(pH-pKa)}\} \qquad (3.24)$$

or

$$C^{aq}{}_{HA} = C_{obs}{}^{aq}/\{1+10^{(pH-pKa)}\} \qquad (3.25)$$

From the definition of the partition coefficient it follows that

$$\Pi = C^{or}{}_{HA}/C^{aq}{}_{HA} = C^{or}/[C_{obs}{}^{aq}/\{1+10^{(pH-pKa)}\}] =$$

$$\{1+10^{(pH-pKa)}\}\{C^{or}/[C_{obs}{}^{aq}\} \qquad (3.26)$$

3.7 PREFORMULATION TASKS

The $pK_{(c)}$ of the compound should be determined by titration or, when possible, by spectrophotometry. If titration is employed then titration of the water should be subtracted (Carstensen, 1972; Parke and Davis, 1954). The dependence of $pK_{(c)}$ on ionic strength should be determined, and from that the pK_a should be determined.

REFERENCES

Brønsted, J.N., (1943a), *Fysisk Kemi*, Munksgaard, Copenhagen, Denmark, p. 273.

Brønsted, J.N., (1943b), *Fysisk Kemi*, Munksgaard, Copenhagen, Denmark, p. 293.

Carstensen, J.T., (1971), *Theory of Pharmaceutical Systems, Volume I*. Academic Press, N.Y., p. 117.

Parke, T. and Davis, W., (1954), *Anal. Chem.*, **25:** 642.

CHAPTER 4

Surface Tension and Critical Micelle Concentration

4.1 SURFACE TENSION

If a ring of radius r (and, hence, circumference $2\pi r$) is placed in the surface of a liquid, then the force that is necessary to remove it, F, defines the surface tension, γ, in the following fashion:

$$F = \gamma 2\pi r \tag{4.1}$$

The unit of surface tension is, therefore, force divided by length, and the unit dyne/cm is often used. By multiplying both sides of Eq. 4.1 with a length dimension, b, it follows that the unit could also be work/area, i.e., erg/cm^2, a unit that is also used frequently.

More attention will be given to surface tension in a later chapter. Let it simply be said at this point, that determining the surface tension of a solution of a drug substance is of importance and should be carried out. It does not, with newer apparatuses such as the Kruss tensiometer and the Cahn tensiometer, require much drug substance. The importance of it lies in the determination of the critical micelle concentration, if indeed the drug substance is an amphiphile.

Further definitions of this will follow in a later chapter and will simply be touched on at this point.

4.2 CRITICAL MICELLE CONCENTRATION

An amphiphile is a molecule that possesses a hydrophilic (water loving) part and a hydrophobic (lyophilic, water-shying) part. The former is often a functional set of groups capable of hydrogen bonding (polyoxyethylene groups, for instance), whereas the latter part is groups such as hydrocarbon chains.

An amphiphile is often presented as a circle (the hydrophilic head) and a wiggly line (the hydrophobic tail, wiggly because it resembles a long hydrocarbon chain). This is depicted in Figure 4.1.

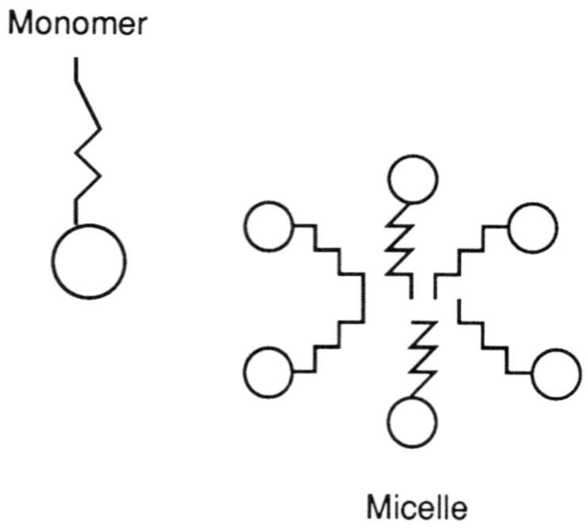

FIGURE 4.1 Schematic of amphiphile and micelle.

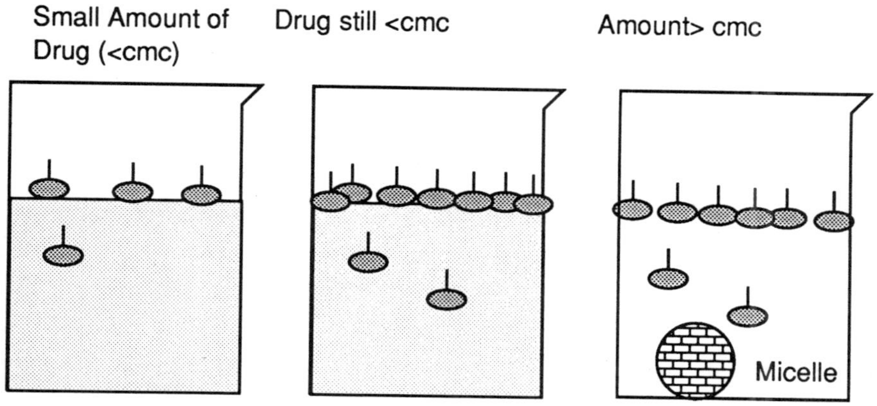

FIGURE 4.2 Formation of micelles by an amphiphile in an aqueous system.

If a drug is an amphiphile, then the following behavior will be observed. Its true solubility, S, is usually fairly low, and by adding small amounts of drug substance to water, S will not be superseded. However, more of the drug substance will be present in the surface layer than in the bulk of the solution.

As more amphiphile is added, the surface "fills" up, and when it is completely filled up, the molecules, when more is added, can only go into the bulk phase or precipitate. In a case of an amphiphile the molecules go into the bulk phase and form micelles.

Micelles, as shown in Figure 4.2, are aggregates of amphiphile molecules where the lyophobic sections group together in a core, and the hydrophilic tails protrude into the aqueous bulk layer, so that hydrophilic groups are where they like to be (in the water) and hydrophobic groups are where they like to be (in the hydrophobic core).

It is noted that, at the point where micelles start forming, there is no more buildup of concentration in the surface, so that from this point on the surface tension should be constant.

On the other hand, before the cmc, the surface tension will keep on decreasing, until the cmc is reached. During this stage it can be shown that the profile follows the so-called Gibbs isotherm equation:

$$\gamma = \beta - RT\Gamma \ln[C] \qquad (4.2)$$

where C is the concentration of added amphiphile and Γ is the so-called surface excess. It is a two-dimensional concentration term (moles/cm^2).

There are not too many reports on drugs that are amphipl Fellmeister et al. (1965) reported on chlorpromazine and Franchini Carstensen published work on a compound, ($C_{26}H_{32}O_4SNa_2$), i.e disodium salt of 2-ß-[(2-carboxyethyl)thio-]-[-8-(2-phenyl)ace benzene propanoic acid:

$R\text{-}CH\text{-}S\text{-}CH_2\text{-}CH_2\text{-}COO^-Na^+$
|
$CH_2COO^-Na^+$

which is an amphiphile. This is referred to as compound I in the following.

Determination of cmc by use of surface tension was reported by Franchini (1995), and data are shown in Figure 4.3.

4.3 EFFECT OF TEMPERATURE ON CMC-VALUES

The value of the cmc often increases (almost linearly) with temperature (Franchini and Carstensen, 1996). An example of this is shown in Figure 4.4.

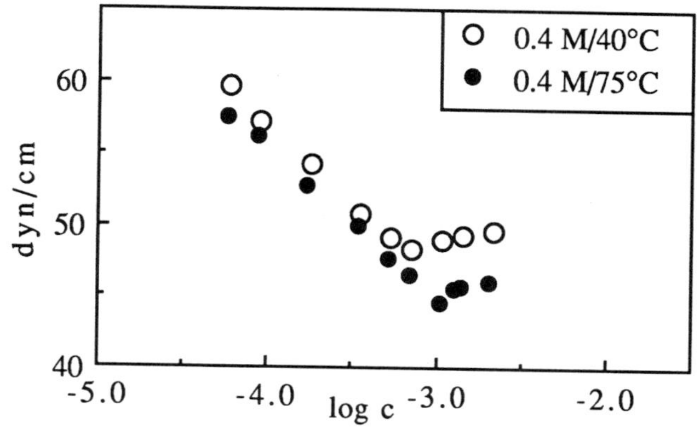

FIGURE 4.3 Surface tension of compound I vs the logarithm of its concentration in aqueous solution at pH 10 in K_2HPO_4/KOH buffer of indicated molarities. Graph constructed from data published by Franchini (1995).

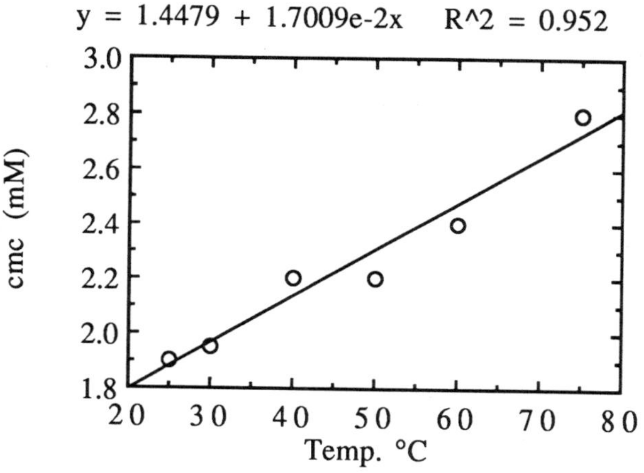

FIGURE 4.4 Effect of temperature on cmc of compound I. Graph constructed from data published by Franchini and Carstensen (1996).

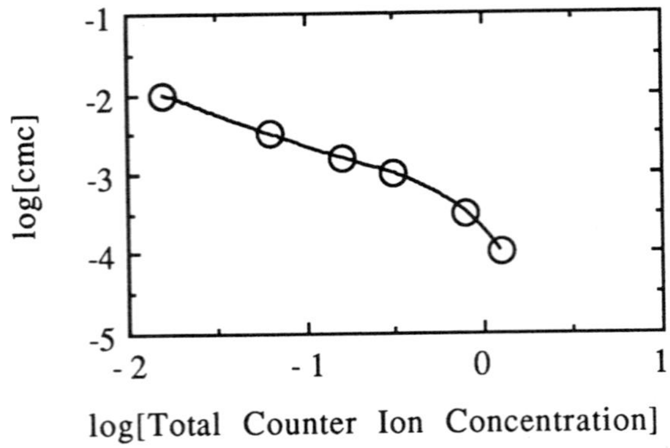

FIGURE 4.5 Cmc of compound I as a function of total counter ion concentration for sodium decyl sulfate at 25°C.

4.4 EFFECT OF COUNTER ION ON SURFACE TENSION

Surface tension is a function of the counter ions in the system. Figure 4.5 shows an example of this.

4.5 EFFECT OF TEMPERATURE ON SURFACE TENSION

The effect of temperature on the surface tension of the compound I described above is shown in Figure 4.6.

4.6 GENERAL IMPACT OF AMPHIPHILIC BEHAVIOR

Although more comments shall be made about amphiphilic behavior in a later chapter, it is important to recognize the amphiphilic nature of a drug substance.

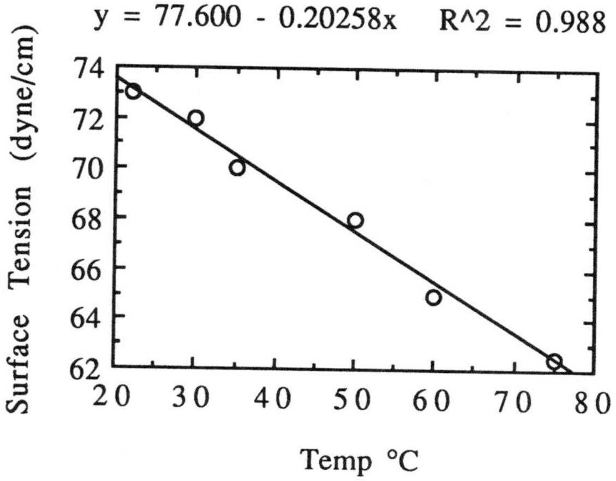

FIGURE 4.6 Surface tension of aqueous solutions of compound I as a function of temperature for a 0.1 M pH 7 KH_2PO_4/KOH buffer.

Amphiphilic behavior may affect kinetics in solution, and it is obvious that even such a fundamental property as a substance adherence to Beer's law is a function of molecular aggregation. By Beer's law

$$\log(I/I_o) = -\varepsilon a C \tag{4.3}$$

a plot of the optical density versus concentration (in dilute solutions) should be linear *and pass through the origin.* In spite of the fact that very often this latter requirement is often not met and there is some obvious, but slight, curvature, it is general practice to draw a straight line and account for the "intercept." The downward curvature is most often due to aggregation (e.g. dimerization), and a more proper procedure would simply be to have a curved calibration curve.

REFERENCES

Fellmeister, A., Schaubman, R. and Howe, H., (1965), *J. Pharm. Sci.,* **54**:1589.

Franchini, M., (1995), *The effect of aggregation state on the degradation kinetics in solution of an oxidizable sulfide dicaboxylic acid drug.* Ph. Thesis, University of Wisconsin, Madison, WI.

Franchini, M. and Carstensen, J.T., (1994) ACS Abstracts of Meetings, National Meeting, March 1994, Part II, PHYS 328.

Franchini, M. and Carstensen, J.T. (1996), *J. Pharm. Sci.,* **85**: 220.

CHAPTER 5

Solution Kinetics

Solution chemistry will always be of importance in preformulation. Development of stability-indicating assays frequently requires that solutions be made and subjected to both accelerated and real-time stability studies.

Before doing this, it facilitates further studies if the decomposition reaction mechanism is known. Most often this is not the case, but the stability-indicating method will indicate that there is a main peak (in HPLC) for the active ingredient, maybe one for the major decomposition product, and a second (set of) peak(s) for "related" compounds.

The mechanisms can be of different types:

(a) straight first order (e.g., hydrolysis)
(b) parallel reactions (also first order)
(c) A—>B—>C (consecutive) reactions
(d) oxidations

In cases (a), (b), or (c) the disappearance of the active[2] will be pseudo-first order, but in the latter case, and in cases of combinations of reaction types, this is not the case.

5.1 PRELIMINARY TASK

Provided an assay exists or has been developed (this is usually the case in the preformulation stage), the task of the preformulator is to establish:

(a) the optimum pH
(b) whether there is a concentration effect
(c) whether there is a specific buffer effect
(d) whether there is a kinetic salt effect

5.2 REACTION ORDER

It is important to establish the reaction order. If only small amounts of drug are present, then one should make solutions that are four times the concentration of the detectable limit once the sample is prepared directly for HPLC. If spectrophotometric assays are used, then one may simply use the solution per se as the solution placed in the cuvette and run spectra at various storage times.

The first task is to determine the reaction order, but if that is not possible at the early time in preformulation, one should assume it to be first order.

Most reactions in solution are pseudo-first order because they are hydrolyses, and as such the drug A will form the hydrolysis product B via the scheme:

$$A + H_2O \dashrightarrow B \qquad (5.1)$$

---

2 The word "active" (rather than "active component") is frequently used.

so the rate equation is:

$$dA/dt = - k_2[A][H_2O] = -(55.5k_2)[A] = -k_1[A] \tag{5.2}$$

where k_2 is the (true) second-order rate constant, where the water concentration is considered constant hence leading to the first-order rate constant:

$$k_2 = 55.5k_1 \tag{5.3}$$

If for instance the solution of A were one molar (much more than is usually the case with drug solutions) then the water concentration would drop from 55.5 to 54.5 if the reaction (5.1) went to completion. Such reactions are called pseudo-first order.

5.3 FRACTIONAL LIVES

Eq. 5.2 integrates to:

$$\ln[A/A_o] = -k_1t = -kt \tag{5.4}$$

where A_o is the initial drug concentration and where the subscript to the rate constant will be dropped unless needed for clarity.

For instance, at the point in time where the concentration has dropped to 95% of its original value, $A/A_o = 0.95$, so that

$$\ln[0.95] = -0.051 = -kt_{0.95} \tag{5.5}$$

$t_{0.95}$ is denoted a fractional life. This can be used, later in the preformulation stage, to establish whether, indeed, a reaction is

TABLE 5.1 Example of First-Order Data

Wk.	Percent Retained			ln[k] ($week^{-1}$)		
	60°	50°	40°C	60°C	50°C	40°C
0	100	100	100	4.605	4.605	4.605
2	80	92	97.5	4.382	4.522	4.580
4	64	85	96	4.159	4.443	4.564

first-order or not. This will be demonstrated at a later point in this text. An example of data that are (presumably) first-order is shown in Table 5.1 and Figure 5.1.

5.4 TEMPERATURE EFFECTS

As manifest from Figure 5.1, a reaction is a function of temperature, and most often the Arrhenius equation holds:

$$\ln[k] = -(E_a^*/RT) + \ln[Z] \tag{5.6}$$

where E_a^* is the activation energy in calories per mole, R is the gas constant (1.99 cal/mole) and T is absolute temperature (°K). For plotting purposes it is more convenient to use 1000/T than 1/T as independent variable, in which case the equation becomes:

$$\ln[k] = -(E_a/R)(1000/T) + \ln[Z] \tag{5.7}$$

where

$$E_a = E_a^*/1000 \tag{5.8}$$

i.e. the energy of activation in such plotting is in kCal/mole.

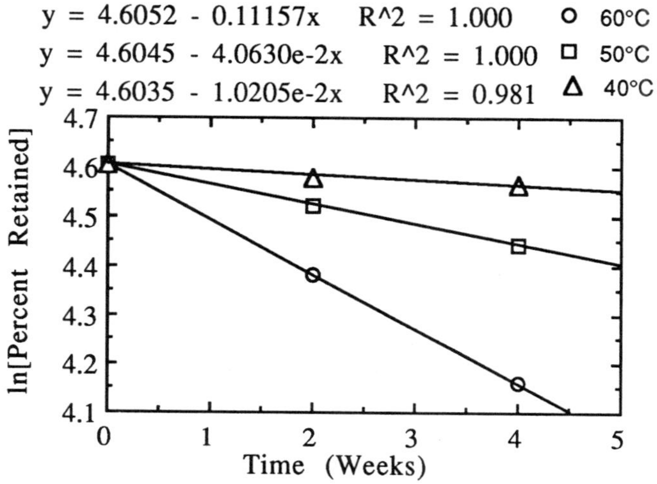

FIGURE 5.1 Decomposition of a drug in aqueous solution. The least squares fits are shown in the inlay.

It is obvious that in such plotting (if not done by a graphics program) it is necessary to convert T (°C) into 1000/T (°K^{-1}). This is best done by programs such as the one shown in Table 5.2.

If, for instance, it is desired to have temperatures from 20 to 80 at 10°C intervals, then one inputs this and obtains the results shown in Table 5.3.

If a graphics program (e.g. Cricket Graph^R) is used then the conversion can be made in the program itself. The Arrhenius plot is used often in preformulation and if so, the data in Table 5.1 serve as a good example.

This means that one may calculate, for example, the rate constant at 30°C (or read it off on the graph) by inserting 1000/T = 3.300 in the least squares equation in Figure 5.2.

TABLE 5.2 Program for Converting from °C to 1000/T

```
INPUT "Highest Temperature=";Y1
INPUT "Lowest Temperature=";Y2
INPUT "Number of Temperatures=";N
Print "Temp.", "1000/T"
Q = (Y1-Y2)/N
FOR T1 = Y2 TO Y1 STEP Q
X1 = T1 + 273.15
X2 = 1000/X1
PRINT T1,X2
NEXT T1
```

$$\ln[k_{30}] = 35.9882 - (12.6931 \times 3.300) = -5.899 \tag{5.9}$$

$$k_{30} = \exp(-5.899) = 0.0027421 \tag{5.10}$$

so that the amount left after 2 years (104 weeks) would be

TABLE 5.3 Output from Program in Table 5.1

Highest Temperature = 80°C
Lowest Temperature = 20°C
Number of Temperatures = 7

Temp.	1000/T
20	3.411
30	3.299
40	3.193
50	3.095
60	3.002
70	2.914
80	2.832

TABLE 5.4 Rate Constants from The Kinetic Data in Table 5.1

Temp,°C	Rate Constant k	ln[k]	1000/T
60	0.115	-2.163	3.002
50	0.0405	-3.206	3.095
40	0.0102	-4.585	3.193

$$C/C_o = \exp(-0.0027421 \times 104) = 0.752 \tag{5.11}$$

or 75%.

It is convenient to have a program to carry out such calculations. One is shown in Table 5.5. When this is done for temperatures of 70 and 80, then rate constants of 0.367 and 1.047 week^{-1} are obtained.

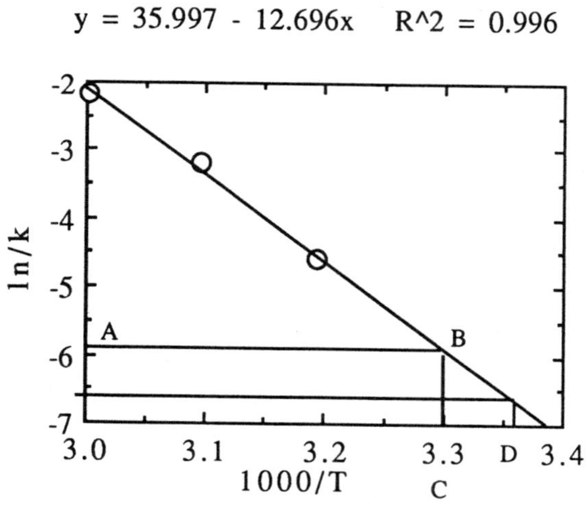

FIGURE 5.2 Data shown in Figure 5.1 treated by the Arrhenius equation. It is seen, graphically, that $\ln[k_{30}]$ = -5.9 and $\ln[k_{25}]$ = -6.6.

TABLE 5.5 Prediction of Rate Constants and Retention

```
INPUT "ln[Collision Factor]=";Z
INPUT "Slope of Arrhenius Plot="; E1
INPUT "Temperature, °C="; T1
T2 = T1+273.15
T3 = 1000/T2
X1 = Z - (E1*T3)
REM X1 is ln[k]
X2 = EXP(X1)
REM X2 is Rate Constant
PRINT "Temperature="; T1
PRINT "Rate Constant=";X2
```

Even though the data in Figure 5.1 have good correlation coefficients and the Arrhenius plot also has a good correlation coefficient, some caution should be exercised in the extrapolations just made.

First of all, it is difficult, from a work-time point of view, to use more than three temperatures, and also to use more than the two or three time points used in Figure 5.1.

The procedure is acceptable if it is used to distinguish between formulations, i.e. extrapolating and selecting the one which is the most stable at room temperature.

It should be noted that if statistical intervaling is used, then the 95% confidence limits on the extrapolated figure are quite wide, i.e., the number is associated with a fair amount of uncertainty.

Finally it should be mentioned that the extrapolation is only valid if the reaction at the higher temperature is the same as that at the lower temperature.

5.5 EFFECT OF REACTION MECHANISMS

If the reaction at the higher temperature is different from the one at the lower temperature or, rather, if there is more than one reaction, then the two reactions will have different heats of activation and one will take over at higher temperatures.

This is an often encountered phenomenon, i.e.

A --> B	k_B and E_B	(5.12)
A --> C	k_C and E_C	(5.13)

where both can be first order, or where one may be first order (hydrolysis) and the other of a different order (zero) because it is a different type of reaction (e.g. oxidation).

<u>Example 5.1</u>

The study in Table 5.1 is extended to 70°C and 75°C with the results in Table 5.6.

It can at times be difficult to distinguish between first and zero order, but the data in Table 5.6 are, indeed, zero order. This refers to the compound I shown in Chapter 4, and for convenience the formula of this compound is repeated below:

Formula: ($C_{26}H_{32}O_4S.Na_2$)

Name: ß-[(2-carboxyethyl)thio-]-2-[-8-(2-phenyl)acetyl] -benzene propanoic acid disodium salt

Formula:

$R\text{-}CH\text{-}S\text{-}CH_2\text{-}CH_2\text{-}COO^-Na^+$
      |
      $CH_2COO^-Na^+$

If this is suspected, then the experiment should be rerun with more time points to, indeed establish the order.

TABLE 5.6 System in Table 5.1 Tested at 70 °C and 75 °C

Time	A/Ao		ln[A/A$_o$]	
	70°C	75°C	70°C	75°C
0	1.0	1.0	0.000	0.000
0.25		0.72		-0.821
0.5		0.44		
1	0.67		-0.400	
2	0.33		-1.109	

But suppose the data were treated first order (Figure 5.3), then the rate constants obtained from least squares fit of the last two columns would be 0.241 and 0.713 week^{-1}. If these two data points were placed in the graph in Figure 5.2, then Figure 5.3 would result.

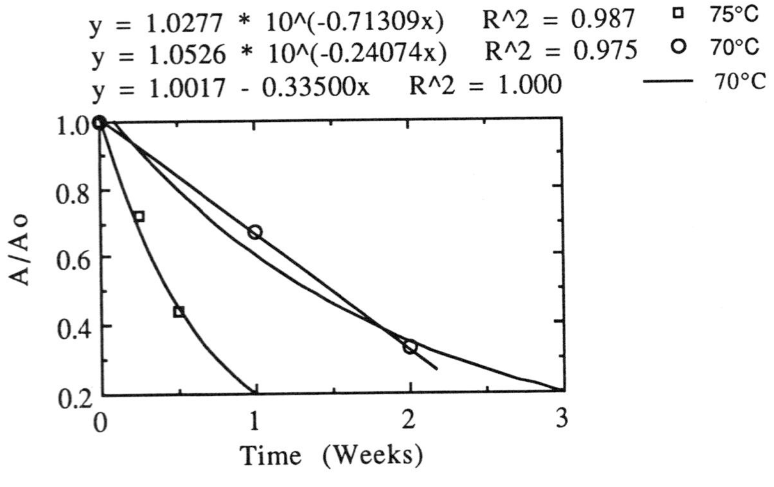

FIGURE 5.3 Data from the example in Table 5.1 obtained at 70° and 75°C. The rate constants will be: 2.303 x 0.241 = 0.555 and 2.303 x 0.713 = 1.642 mo^{-1}. Also shown are the 70°C data treated as a zero-order reaction, and it is seen that the fit is much better than the first-order fit.

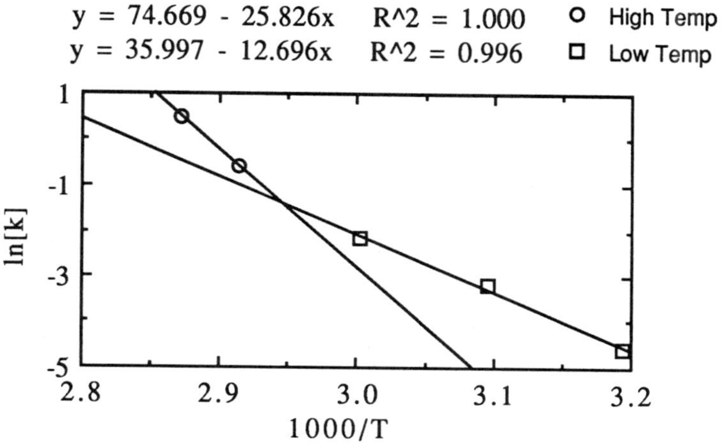

FIGURE 5.4 Arrhenius plot of parallel reactions.

First of all, note that the points actually lie on a straight line, i.e. that they are zero order, yet we have treated them by exponential decay (first order). If we do these separately on an Arrhenius plot, then there will be two graphs which intersect. One reaction (erroneously treated as first order in Figure 5.3) is faster at the higher temperature, the other is faster at the lower temperature.

The danger here is that, suppose we had done the Arrhenius plot at 50°C, 60°C and 70°C, we would then have obtained the data in Table 5.7.

TABLE 5.7 Data from 50, 60 and 70°C Only

Temp,°C	Rate Constant k	ln[k]	1000/T
70	0.555	-0.589	2.914
60	0.115	-2.163	3.002
50	0.0405	-3.206	3.095

The data in Table 5.7 are plotted in Figure 5.5, and the least squares fit equation is shown. On the surface the fit is good (R = 0.99), but comparing it with Figure 5.4 it is obviously incorrect.

If the data in Figure 5.5 are extrapolated to 25°C (1000/T = 3.355), then it is found that:

$$\ln[k_{25}] = 41.3505 - (14.4279 \times 3.355) =$$

$$41.3505 - 48.4056 = -7.0551 \qquad (5.14)$$

Hence

$$k_{25} = \exp(-7.0551) = 0.000863 \text{ week}^{-1} \qquad (5.15)$$

After 104 weeks the fraction retained would be

$$\exp(-0.000863 \times 104) = e^{-0.9} = 0.914 \qquad (5.16)$$

i.e., 91.4% would appear to remain. In contrast to this, if only the lower temperatures were used, the amount would be (Figure 5.2):

$$\ln[k_{25}] = -0.6 \text{ so } k_{25} = 0.00136 \text{ wk}^{-1} \qquad (5.17)$$

i.e. after 104 weeks the fraction retained would be:

$$A/A_o = \exp(-0.00136 \times 104) = 0.868 \qquad (5.18)$$

or 87%, i.e. less than predicted in the incorrect fashion.

This type of error is frequently committed. It is assumed that the fact that the fit is not perfect is attributable to errors, but the best way of testing for this is to test at more temperatures.

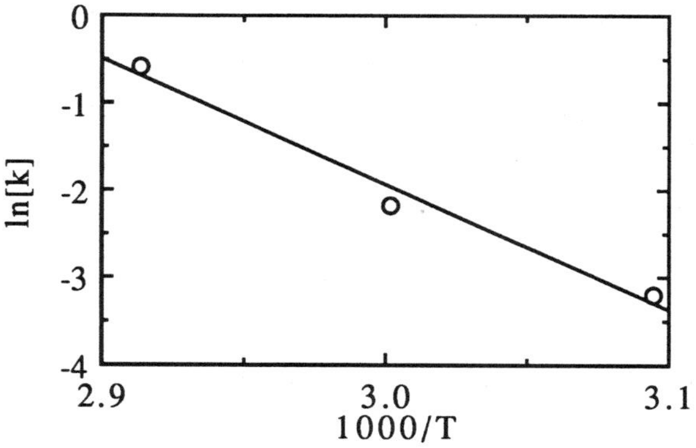

FIGURE 5.5 Apparent Arrhenius plot of the data in Table 5.7. The least squares fit equation is y = 41.3505 - 14.4279x.

5.6 CALORIMETRIC METHODS

The issue then arises, in preformulation, that Arrhenius plotting is not necessarily a good measure from which to extrapolate unless it is carried out at quite a few (e.g. five) temperatures. If all five temperatures had been tested in the examples above, then a fairly accurate and precise assessment of room temperature stability *in solution* could be obtained.

The reason the words *in solution* are emphasized is that the content uniformity is not a problem in solutions, so that, for all practical purposes, only the assay variation contributes to the precision of the rate constants. The lower the temperature, the more pronounced this gets, for example if the loss of drug is less than the sensitivity of the assay. In other words, at room temperature, one must keep the product much longer to see "meaningful" losses.

There are, today, very highly insulated, high-precision calorimeters, where one may place a solution for, say, thirty days at 25°C. There will be some initial spurious heat events, after which an "equilibrium" will start. Here there will be a heat evolution which is due to decomposition, and hence can be plotted. If this is done at a higher temperature, where a loss can be determined chemically after a given time, then the heat event can be quantitated in terms of amount decomposed, and an accurate[3] and precise kinetic profile can be obtained at room temperature.

5.7 pH PROFILES AND BUFFER EFFECT

Solution kinetics are usually affected by pH, and to investigate this, it is customary to check the stability (determine the rate constants) at different pH-values. This is done by use of buffers.

The rate constants, k, are usually a function of both $[H^+]$, $[OH^-]$, $[B^-]$ and [HB] where the latter denotes an un-ionized and ionized buffer species (e.g. Maulding et al., 1975). These are usually citrate or phosphate.

The dependence is usually cast as:

$$k = k_o + k_+[H^+] + k_-[OH^-] + k_1[B^-] + k_2[HB] \quad (5.19)$$

where the subscripts denote where the rate constant dependence derives from and the terms in brackets are concentrations. k_o is an inherent rate constant that is independent of the concentrations of other species than the drug itself (if that).

At low pH, the term $k_-[OH^-]$ falls out and at a given pH the ratio of

$$[B^-]/[HB] = 10^{-(pH-pKa)} \quad (5.20)$$

so we may think of the buffer concentration [Bu] as being the determining factor.

---

[3] This, again, depends on the assumption that the reaction at 40°C is the same as the one at 25°C.

5.7.1 BUFFER EFFECT

A buffer, such as a phosphate buffer, has a certain pK_a (in fact has three), so that, if it is present in solution, then production of acid or alkaline decomposition products will not change the pH too much, because of the presence of the buffer. The effectiveness of the buffer is denoted its buffer capacity and is defined as

$$\beta = d(pH)/d(Q) \tag{5.21}$$

where Q is the moles of either acid or base produced in the decomposition.

The drug substance in Table 5.1 refers to a pH of 4, obtained with a phosphate buffer in a concentration of 0.01 M. (The 40°C data are repeated for convenience in Table 5.8.) If this experiment is carried out in different buffer concentrations, then the results in Table 5.8 result.

It is seen from Eq. 5.19 that when the hydroxyl ion concentration term is neglected and the buffer concentration is studied, then

$$k = k_o + k_+[H^+] + k[Bu] \tag{5.22}$$

TABLE 5.8 Effect at pH 4 Carried of Different pH-4-Buffers

Wk.	Percent Retained			ln[k] (week^{-1})		
	0.01M	0.02M	0.04M	0.01M	0.02M	0.03M
0	1.00	1.00	1.00	0	0	0
4	0.981*	0.974	0.971	-0.019	-0.025	-0.029
8	0.961	0.951	0.946	-0.040	-0.050	-0.056
k (week^{-1})				0.0050	0.0062	0.0069

*Precision is exaggerated in this example to illustrate a point.

But if the pH is constant, then the first two terms are constant, so that

$$k = \{k_o + k_+[H^+]\} + k[Bu] \qquad (5.23)$$

and the term $\{k_o + k_+[H^+]\}$ can be obtained by plotting k as a function of buffer concentration and extrapolating to zero. The data from Table 5.8 are plotted in this fashion in Figure 5.6.

It is seen, both from Table 5.8 and Figure 5.6 that there is an effect of buffer concentration, and this is shown graphically in Figure 5.7.

It is noted that at zero buffer concentration the value of k is 0.0082 weeks^{-1}, i.e.

$$\{k_o + k_+[H^+]\} = 0.0041 \qquad (5.24)$$

at pH 4.

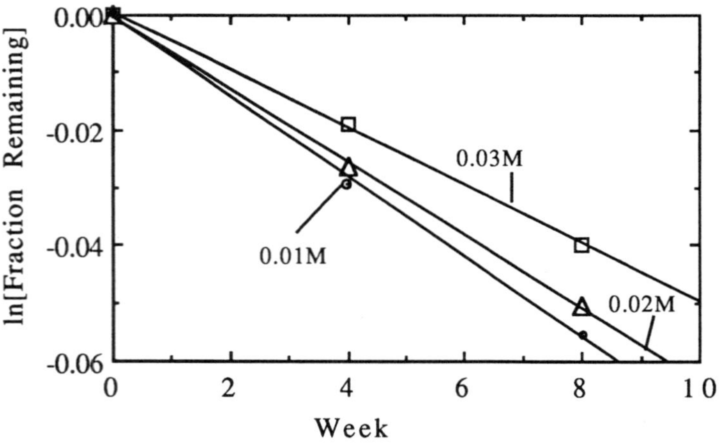

FIGURE 5.6 Data from Table 5.8.

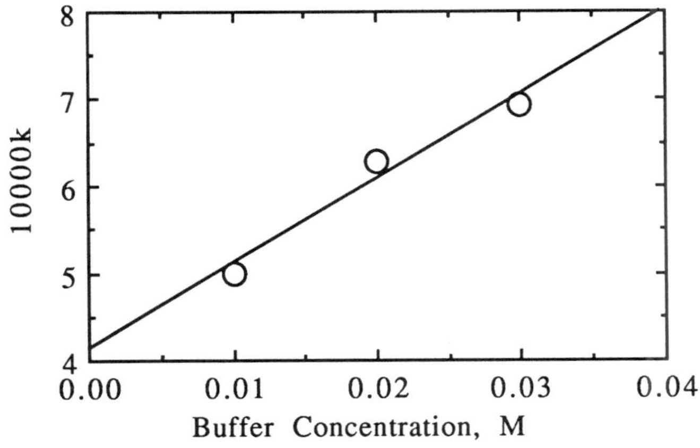

FIGURE 5.7 Data from Figure 5.6.

5.7.2 KINETIC pH PROFILES

It is noted (Carstensen, 1972; Carstensen, 1995) that if this is carried out at different low pH values, then the second term in Eq. 5.23 becomes predominant, and the equation becomes:

$$k = k_+[H^+] \quad (5.25)$$

or

$$\log_{10}[k] = \log_{10}[k_+] + \log[H^+] = \log_{10}[k] - pH \quad (5.26)$$

If, therefore, the experiment above is carried out at several low pH values, then a plot of $\log_{10}[k]$ versus pH will be a straight line with a slope of negative one.

Similarly, at high pH

$$\log_{10}[k] = \log_{10}[k_-] + \log[OH^-] \quad (5.27)$$

but, as shown previously:

$$[H^+][OH^-] = 10^{-14} \tag{5.28}$$

or

$$\log[OH^-] = -14 - \log[H^+] = pH-14 \tag{5.29}$$

so that Eq. 5.27 becomes

$$\log_{10}[k] = \log_{10}[k_-] + pH - 14 \tag{5.30}$$

Hence, in this high pH-region, the pH-profile should be a straight line with a slope of positive one.

In areas where k_o is larger than both $k_+[H^+]$ and $k_-[OH^-]$ the profile becomes "flat," so that a common pH-profile will have the appearance shown in Figure 5.8.

In Figure 5.8, the line segment AB corresponds to Eq. 5.26 and DE to Eq. 5.30. The segment CD is the area where k_o is the predominant term in Eq. 5.19.

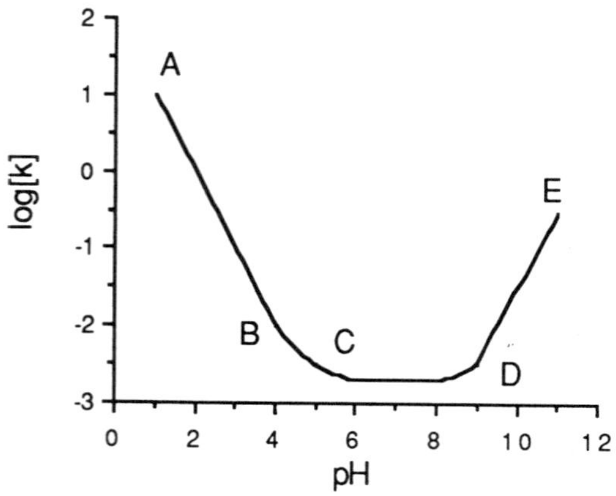

FIGURE 5.8 Common pH profile.

5.8 OXIDATION

The previous writing has been concerned with *hydrolysis*. At times, however, a substance will be prone to oxidation, and in this case the kinetic profile can be of two types: (a) auto-oxidative, or (b) plain oxidative.

5.8.1 AUTO-OXIDATION

Without touching on mechanisms it should be said, up front, that the worst offender in oxidation is the presence of metal ions, and that the best remedy in such cases is a metal chelator, such as ethylenediamine tetraacetic acid (EDTA). It is usually obtained as the disodium salt and is most effective in ranges from pH 5 to 8.

The two first columns in thefollowing set of data (Table 5.9) are characteristic of an oxidative decomposition profile. The data are plotted in Figure 5.9. Frequently such data fit the equation:

$$\ln[x/(1-x)] = -k(t-t_i) \qquad (5.31)$$

The data in Table 5.9 are treated in this fashion and are shown in Figure 5.10.

Often the last points do not fit well. As mentioned, in the case of auto-oxidation, the presence of metal ions is deleterious, and studies should be carried out to ascertain whether this is so or not (Thakur et al., 1991). The use of ethylenediamine tetraacetic acid is advocated in such cases.

TABLE 5.9 Auto-Oxidative Decomposition Data

Weeks	x	(1-x)	x/(1-x)	ln[x/(1-x)
0	1			
3	0.99	0.01	99.000	4.595
6	0.98	0.02	49.000	3.892
9	0.96	0.04	24.000	3.178
12	0.88	0.12	7.333	1.992
15	0.7	0.3	2.333	0.847
18	0.55	0.45	1.222	0.200
21	0.4	0.6	0.667	-0.405
24	0.32	0.68	0.471	-0.753
27	0.3	0.7	0.429	-0.846
30	0.28	0.72	0.389	-0.944

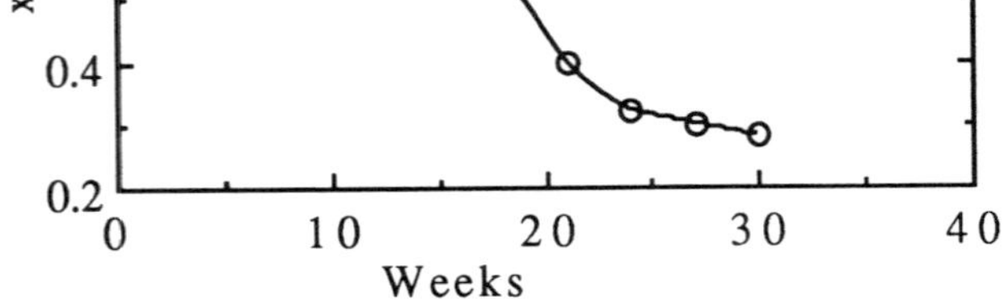

FIGURE 5.9 Data from Table 5.9.

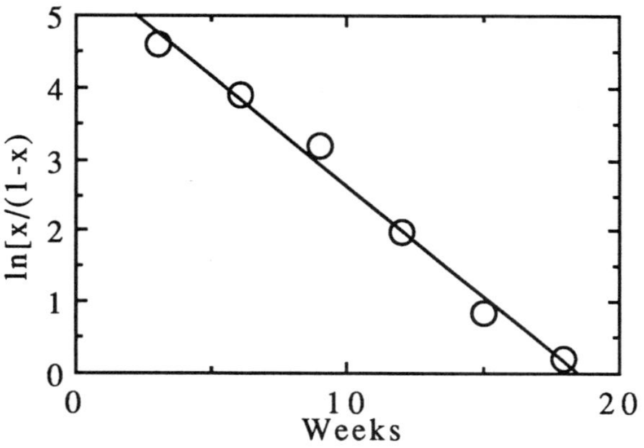

FIGURE 5.10 Data from Table 5.9 treated by Eq. 5.31. The last three points have been omitted.

5.8.2 PLAIN OXIDATION

As mentioned the rote cause or catalyst for auto-oxidation is metal ions and free radicals. Often EDTA will complex the metals, and the oxidation then becomes plain.

The amount of oxygen (C_{O2}) in a container distributes itself between head space and solution by Henry's law:

$$C_{O2}/P_{O2} = \text{Constant} \tag{5.32}$$

where P_{O2} is the oxygen pressure in the head space, usually 0.22 atm. This amount is usually in great excess, molarly, of the drug in solution. This latter reacts with oxygen in solution, and when an oxygen molecule is used up according to:

$$A + O_2 \text{ --> Decomposition} \tag{5.33}$$

then it is replaced by one from the head space. Hence the second-order equation dictated by Eq. 5.33:

$$dA/dt = -k_2[A][O_2] \tag{5.34}$$

becomes (approximately)

$$dA/dt = -k[A] \tag{5.35}$$

where

$$k = k_2[A] \tag{5.36}$$

is almost constant. Such equations, then, become pseudo-first order, since Eq. 5.36 is a first-order equation.

The effect of temperature on oxidation reactions is such that the rate constant in Eq. 5.34 increases with increasing temperature, but the fact that oxygen is less soluble at higher temperature makes the overall reaction rate slower at a given high temperature, so that the Arrhenius equation per se is not valid (Franchini and Carstensen, 1994).

Salts also affect the oxygen solubility, so that they influence the reaction rate given by Eq. 5.34 in a manner over and above the kinetic salt effect, next to be discussed (Franchini et al., 1993).

Finally, antioxidants should always be tested when oxidations are possible. Some antioxidants work by inhibiting either the formation or reaction of free radicals, and others act simply as scavengers. As a case of the latter, Pudipeddi et al. (1992) have shown that in the case of procainamide, sodium bisulfite was effective until used up.

5.9 KINETIC SALT EFFECT

Many reactions involve ionic species, and when this happens, the rate constant is a function of the ionic strength, μ, of the system (Carstensen, 1970).

If the reaction is:

A + Q --> Decomposition

then the salt effect is:

$$\log_{10}[k/k_o] = -Z_A Z_b[\sqrt{\mu}/(1+\sqrt{\mu})] = -\ Z_A Z_b ß \qquad (5.37)$$

where

$$ß = [\sqrt{\mu}/(1+\sqrt{\mu})] \qquad (5.38)$$

and where Z is the charge of the ion in question. If the drug, A decomposes primarily as the anion (A^-) reacting with either H^+ or OH^-, then there will be an ionic strength effect. This should always be tested for in preformulation.

Table 5.10 shows the rate constants for a first-order hydrolysis carried out at a series of ionic strengths. The rate constants, k, have been divided by the rate constant in zero ionic strength solutions.

The data in Table 5.10 are plotted in Figure 5.11. It is noted that the slope is -0.8. Eq. 5.37 would seem to predict that the slope should be an integer. This would be the case if there were only one species of the drug in solution, but if, for instance, the solution were at the pK of an acid (HA), then half of it would

TABLE 5.10 Effect of Ionic Strength on Rate Constants

μ	ß	k/k_o	$\log[k/k_o]$
O.1	0.24	0.643	-0.192
0.2	0.31	0.565	-0.248
0.5	0.41	0470	-0.328
0.7	0.46	0.429	-0.368
0.9	0.49	0.406	-0.392

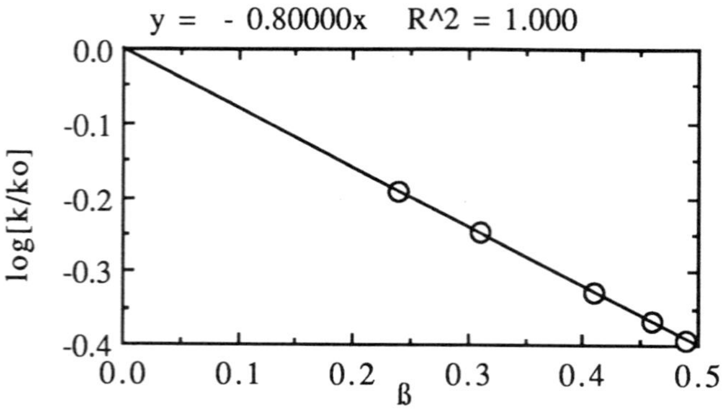

FIGURE 5.11 Data from Table 5.10.

be undissociated HA and not subject to the kinetic salt effect. A^- might be, and this is the reason why fractional coefficients often are the result of kinetic salt effect studies. What is important is that there is or is not an effect.

In the absence of an effect reactions such as $A^- + OH^-$ may be eliminated from consideration in the reaction mechanism scheme.

5.10 AMPHIPHILIC DRUGS

Compound I shows an interesting characteristic, namely that the reaction mechanism in aqueous solution shifts about 75°C (Carstensen and Franchini, 1994; Franchini and Carstensen, 1994; Franchini, 1995). The reason for this is that it may hydrolyze and then oxidize (which happens at high temperature), or oxidize and then hydrolyze (which happens at lower temperature).

The reason for this is the following. The reaction steps are, in rough terms:

$$I + O_2 \rightarrow B_1 \tag{5.39}$$

$$dI/dt = -k_2[O_2][I] \quad \text{(rate constant } k_2) \tag{5.40}$$

$$I + H_2O \rightarrow B_2 \tag{5.41}$$

$$dI/dt = k_1[I] \quad \text{(rate constant } k_1) \tag{5.42}$$

The values of k increase (by an Arrhenius equation) as temperature increases, but $[O_2]$, the oxygen concentration *in solution* decreases with increasing temperature, and at 75°C the values of k_2, k_1 and $[O_2]$ become such that

$$k_1[I] > k_2[O_2][I] \tag{5.43}$$

Hence, at temperature higher than 75°C (in the case of compound I) the hydrolysis becomes the predominant reaction. Data at 93°C are shown in Figure 5.12.

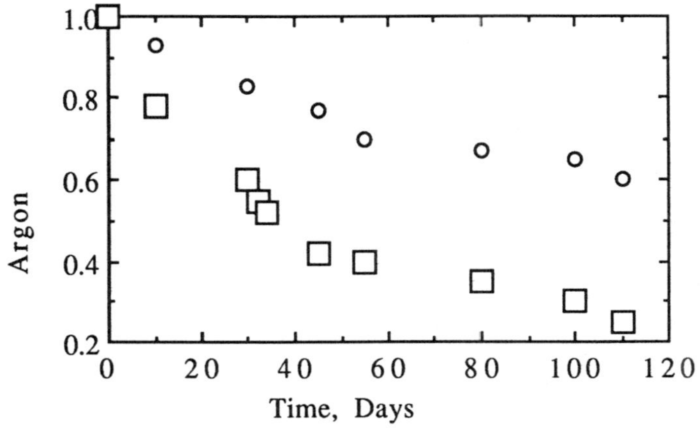

FIGURE 5.12 Decomposition of compound I at 93°C below the cmc. All data are in 0.1 M buffer. Circles are under air, and squares are under argon (i.e. anaerobic). Graph constructed from data published by Franchini (1995).

The question, in such a case, is whether the reaction mechanism under all the conditions (oxygen, anaerobic, etc.) is the same. The half lives of all the situations encountered experimentally are shown in Table 5.11.

If all the time points are divided by their respective half-life, then the points all fall on a common curve, showing that they all adhere to the same reaction mechanism.

It is seen, therefore, that the points fall on the same curve, and that, therefore, the mechanism is the same regardless of the conditions at 93°C.

The actual mechanism may be divined by the shape of the curve in Figure 5.13. If A_1 denotes a monomer molecule in the bulk phase, and A_i in general an i-mer, then there are a series of possible equilibriae. If $i > 3$ is disregarded, then:

$A_1 + A_1 \leftrightarrow A_2$ Equilibrium Constant K_1 (5.44)

$A_1 + A_2 \leftrightarrow A_3$ Equilibrium Constant K_2 (5.45)

The observed amount of drug, A_o, is given by:

$$A_o = A_1 + A_2/2 + (A_3/3) \quad (5.46)$$

TABLE 5.11 Half-Lives Below the Critical Micelle Concentration

Condition	Half-life (days)
Argon, pH 8, 93°C	120.3±15.3
Air, pH 8, 93°C	119.7±6.8
Oxygen, pH 8, 93°C	58.3±5.8
Argon, pH 8, 75°C	437±127
Air, pH 8, 75°C	478±119
Oxygen, pH 8, 75°C	269±98
Argon, pH 8, 60°C	7660±5000
Air, pH 8, 60°C	1585±356
Oxygen, pH 8, 60°C	733±193

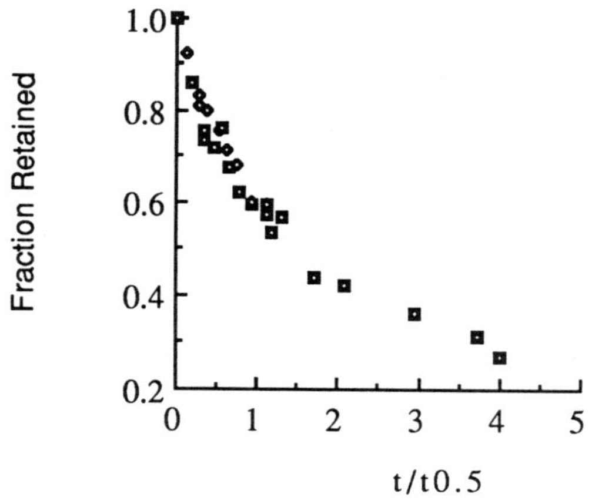

FIGURE 5.13 Data from Figure 5.11. Decomposition of compound I in aqueous solution, presented in a form that shows that the mechanism is the same at the following conditions: All data are in 0.1 M buffer. Circles represent a head space of air, triangles the precence of argon (i.e. anaerobic conditions) and squares the presence of oxygen. For each curve, the real times have been divided by the half time of that particular experiment. All the points are seen to fall on the same curve.

Eq. 5.44 is applied to give Eq. 5.47.

$$[A_1]^2 = A_2K_1 \tag{5.47}$$

or

$$[A_1] = \{[A_2]K_1\}^{1/2} \tag{5.48}$$

Furthermore,

$$A_3 = K_2[A_1][A_2] = K_2[A_2]^{3/2}K_1^{1/2} \tag{5.49}$$

where Eqs. 5.47 and 5.48 have been used in the last step.
Denoting

$$K_2^{1/2}/2= \text{ß}^{1/2} \tag{5.50}$$

and

$$K_2K_1^{1/2}/3= \text{ø}^{1/2} \tag{5.51}$$

Combining Eqs. 5.46, 5.48, 5.49, 5.40 and 5.51 gives:

$$A_o= \text{ß}A_2^{1/2} + A_2 + \text{ø}A_2^{3/2} \tag{5.52}$$

or rearranging:

$$(A_o-A_2)^2 = \{\text{ß}A_2^{1/2} + \text{ø}A_2^{3/2}\}^2 = \text{ß}A_2 + \text{ø}A_2^3 + 2\text{ßø}A_2^2 \tag{5.53}$$

Expanding the left-hand side and rearranging then gives

$$A_o^2 - 2A_2A_o - \{(\text{ß}+ 2\text{ßø}-1)A_2^2 +\text{ø}A_2^3\} = 0 \tag{5.54}$$

This is now solved for A_o:

$$A_o = A_2 \pm \{A_2^2 + [(\text{ß}+ 2\text{ßø}-1)A_2^2 +\text{ø}A_2^3]\}^{1/2} =$$

$$A_2 \pm \{\text{ø}A_2^3 + (\text{ß}+ 2\text{ßø})A_2^2 \}^{1/2} =$$

$$A_2 \{1\pm \{\text{ø}A_2 + (\text{ß}+ 2\text{ßø})^{1/2}\}\} \tag{5.55}$$

In ranges where

$$\text{ø}A_2 >> (\text{ß}+ 2\text{ßø})^{1/2} \tag{5.56}$$

The expression for A_o becomes:

$$A_o = A_2 + \phi^{1/2} A_2^{1/2} \tag{5.57}$$

If A_2 decomposes by first order, i.e.

$$A_2 = A_{2(0)} e^{-kt} \tag{5.58}$$

then Eq. 5.57 becomes

$$A_o = A_{2(0)} e^{-kt} + A_{2(0)}^{1/2} e^{-kt/2} \tag{5.59}$$

This is exactly the pattern seen in Figure 5.14. The data have been treated in this figure by non-linear curve fitting (Sigma-Plot), using a first estimate of A = 0.35 and that of a= 0.001 yielding as a best fit

$$y = Ae^{-at} + (\sqrt{A})e^{-at/2} \tag{5.60}$$

or

$$y = 0.384e^{-0.9t} + 0.62e^{-0.45t} \tag{5.61}$$

where t is reduced time.

However, if the kinetics are carried out in the same manner as shown in Figure 5.11, but *above the cmc, then the reaction becomes zero order.*

When a point in time is arrived at, where sufficient drug is available, it is, therefore, important to *carry out kinetics at as high concentrations as possible. Higher concentrations may stabilize and may destabilize the drug substance.*

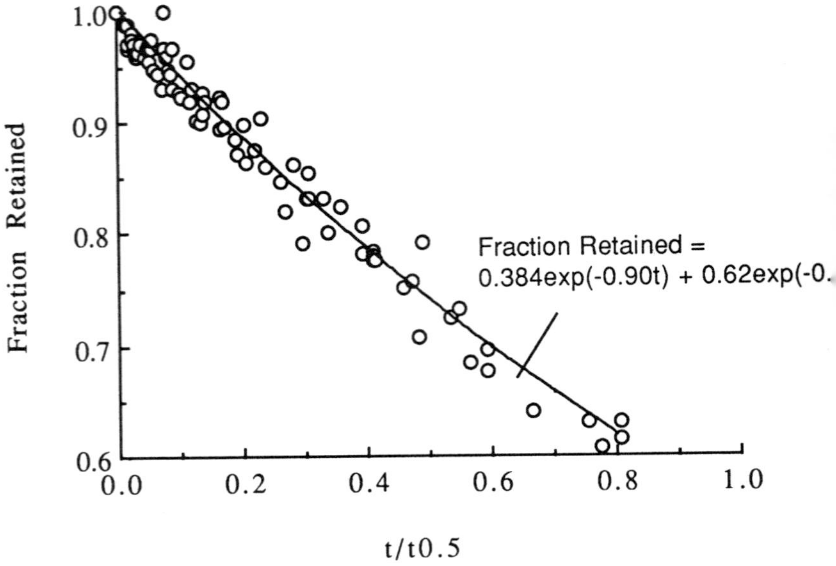

FIGURE 5.14 All data from experiments below the cmc plotted via Eq. 5.57. Graph constructed from data by Carstensen and Franchini (1994) and Franchini (1995).

5.11 OPTIMIZATION

In the early stages of development it is desirable, in as few experiments as possible, to determine what is the optimum buffer, what the optimum buffer concentration is, whether there is a kinetic salt effect and what the best pH is for solutions of the compound. For these reasons, solutions are made up at, e.g. 4 different pH's, at two different buffer concentrations each, and two different salt concentrations each.

In the early stages the concentrations of drug used should be as small as possible, because of the scarcity of drug An adequate amount would be 10^{-2} times the level of detection by, presumably, the HPLC method used. However, in later development it is important to carry out the study at higher concentrations, particularly if it is found that there is a possibility that the drug is amphiphilic. The buffers of preference are phosphate or citrate buffers if it is anticipated that an oral liquid should be developed, but if it is a question of a parenteral product, then an acetate buffer may also be considered. This, however, has a narrower range.

The solutions are then tested at three different temperatures (T°K).

Arrhenius plots are carried out of each preparation, and the pH-profile constructed for each temperature at (extrapolated) k-values at zero buffer concentration.

The slopes of each buffer graph (for the two buffers) are compared, and the most favorable buffer selected.

The optimum pH (the line segment CD in Figure 5.8) is plotted as a function of temperature and "extrapolated" to room temperature, so that an estimate of the optimum pH range is obtained at 25°C.

With the above information, a formulator can start formulating a solution product with the expectation of obtaining the best formula with the smallest amount of experimentation. More to the point, maybe, is that *knowledge of solution kinetics is necessary in order to rationally assess solid state stability.* Since magnesium stearate is often used in solid dosage forms, a separate kinetic study in dilute solutions of magnesium chloride at the "natural" pH of the drug substance would be in order.

5.12 PREFORMULATION CHECK SHEET

In view of the preceding writing a check sheet can be made to ascertain that as many of the basic pieces of information as possible have been checked.

What is the pK_a of the compound?
What is the solubility in water?
What is the pH of a ..... mg/mL[4] solution in water?
What is the solubility-pH curve?
If applicable[5] what is the solubility in
 10 and 20% polyethylene glycol?
 glycerin?
 ethanol?
 50% sugar syrup?
What is optimum pH at 25°C?
What is the reaction order?
Is there oxidation (with and without EDTA)?
What is the activation energy (range)?
What is the best rate constant (or t_{90}) at 25°C?
What is the best projected % retained after 2 years at 25°C?
Is there a kinetic salt effect? If so how serious.

REFERENCES

Carstensen, J.T., (1970), *J. Pharm. Sci.*, **59**:670.

Carstensen, J.T. (1972), *Theory of Pharmaceutical Systems,* I, Academic Press, New York, p. 115.

Carstensen, J.T., (1995), *Drug Stability*, Marcel Dekker, New York, pp. 17-122.

Carstensen, J.T. and Franchini, M.K., (1994), *Pharm. Res.*, **11S**: 267.

Franchini, M.K., Unvala, H. and Carstensen, J.T., (1993), *J. Pharm. Sci.,* **82**:5.

---

4 Specify concentration, since the pH depends on concentration.
5 This depends on whether a formulation with solvents is necessary.

Franchini, M.K. and Carstensen, J.T., (1994), *Int. J. Pharmaceutics*, **111**:153.

Franchini, M.K., (1995), *Effect of aggregation state on degradation kinetics in solution of an oxidizable sulfide dicarboxylic acid drug*. Ph.D. Thesis, University of Wisconsin.

Maulding, H.V., Nazareno, J.P., Pearson, J.E. and Michelis, A.F., (1975), *J. Pharm. Sci.*, **64**:278.

Pudipeddi, M., Alexander, K., Parker, G.A. and Carstensen, J.T., (1992), *Drug Dev. Ind. Pharm.*, **18**:2135.

Thakur, A.B., Morris, K.R., Jerzewski, R.E.L. Wadke, D.A. and Carstensen, J.T., (1991), *Pharm. Res.*, **8**:55.

CHAPTER 6

Suspensions

From the point of view of pharmaceutical systems, it is rare that a suspension is the first dosage form thought of. For this reason, the limitations mentioned earlier, regarding the small amounts of drug substance available, often are not severe problems when it comes to formulating suspensions, because this type of formulation occurs later in the development of novel products. Preformulation is necessary, but many of the micromethods necessary in ordinary preformulation are not as pertinent here.

6.1 DEFINITION

A suspension is a system with two or more phases with at least one solid phase and one liquid phase. In other words, the solubility of at least one of the solutes is superseded. For simplicity the first part of the discussion to follow is confined to a system of two compounds, A and B, one of which (A) is a liquid, and the other (B) is a solid in the temperature range in question.

This latter is usually between 5°C and 50°C for pharmaceutical considerations.

6.2 SOLUBILITY

If a compound has a solubility of C_s g/g of solvent, and if the respective densities are ρ_1 and ρ_2 then C_s g of solid occupies (C_s/ρ_1) mL and 1.0 g of solvent occupies $(1/\rho_2)$, hence the saturation concentration on a volumetric basis (Denoted S) is:

$$S = C_s/[(C_s/\rho_1) + (1/\rho_2)] \tag{6.1}$$

The advantage of expressing solubilities as C_s is that the mass units in this are not a function of temperature. The solubility, of course, is a function of temperature and is, generally, given by:

$$\ln[C_s] = \beta - \Delta H/(RT) \tag{6.2}$$

ΔH is here the heat of solution (which is assumed to be independent of temperature), R is the gas constant, and T is the absolute temperature. Given, for instance, the data in columns 1 and 4 in Table 6.1, the conversions shown in the remaining columns can be arrived at. These data are plotted according to Eq. 6.2 in Figure 6.1.

The more common concentration unit is S (mg/mL), and in that case a "less correct" relation is arrived at, namely:

$$\ln[S] = q - \Delta H'/(RT) \tag{6.3}$$

TABLE 6.1 Example of Solubility Versus Temperature Data

t°C	T°K	1000/T	C_s, g/g	ln[C_s]
5.000	278.150	3.595	1.105	0.100
25.000	298.150	3.354	1.649	0.500
40.000	313.150	3.193	2.203	0.790
50.000	323.150	3.095	2.718	1.000

where q and ΔH' are constants, and many reported solubility curves are plotted in this fashion. This may result in curvature when data are plotted according to Eq. 6.3. However, certain systems exhibit curvature when ΔH is not independent of temperature, or when activity coefficients are dependent of temperature and in such cases, plotting according to Eq. 6.2 leads to curvature as well. If the solubility is plotted as mg/mL, then the value of ΔH' is not truly the heat of solution.

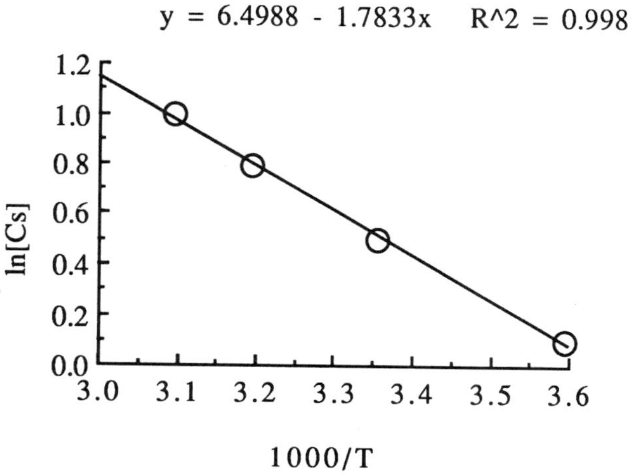

FIGURE 6.1 Data from Table 6.1.

6.3 RATIONALE FOR SUSPENSIONS

There are several reasons for a company wanting to market a product as a suspension. The most obvious one is that the drug is insoluble and cannot be made soluble (and remain therapeutically active) by conventional means. A typical example of this is antacids. One could obviously get them into solution by acidifying them, but then they would no longer be antacids.

6.4 TASTE

One reason is *taste*. If a drug is very foul tasting (mostly bitter), part of the taste response is due to the drug in solution in the saliva after administration. If a solution is administered (e.g., containing 100 mg/mL), then the concentration in the saliva after the liquid and saliva mixes will be high (e.g. 10 mg/mL), and a taste response is recorded.

If the compound is very insoluble (e.g. 0.01 mg/mL), then the concentration in the saliva after mixing with the administered drug substance will also be 0.01 mg/mL, and if this is below the taste threshold, then the suspension will appear to be "tasteless."

An example of this is Tetracycline Base suspensions. The hydrochloride is very soluble and very bitter, the base, however, has a solubility below the threshold of taste.

If a deliberate program is instituted to make a product free of taste by developing poorly soluble salts, a taste threshold profile should be developed by establishing a taste panel. This panel is asked to taste highly diluted solutions of the compound, and then higher and higher concentrations until the threshold is arrived at where taste is detected. For instance as shown in Table 6.2 four testers have tested a solution of a bitter drug in successively increasing concentrations. From the data it is

TABLE 6.2 Example of Taste Panel Results

Concentration	Taste Response A	B	C	D
0.001	-	-	-	-
0.01	-	-	-	-
0.1	+	-	+	-
0.2	+	+	+	+

concluded that the taste threshold is 0.1 mg, and that for a derivative of the drug to be void of taste it must have a solubility of less than that.

A company should always have a taste panel, which has been set up by selection. The tasters (A-D) should preferably have the same sensitivity. This could be done by selecting a larger group of people and subjecting them to a test like the one above, using various compounds of various degrees of bitterness (quinine down to saccharine, for instance), sweetness (sugar to saccharine) and sourness (citric acid to lactic acid).

6.5 STABILITY

Another situation calling for employing a suspension as a dosage form is when a drug in solution is unstable. The general theory is that for suspensions the solid phase is stable, and the solution decomposes. [It should be noted that this assumption is not completely necessary, and that the stability of the solid phase may be accounted for as well (Carstensen, 1995)]. Making the assumption that only solution decomposition matters then, *in* a solution, the decomposition is given by:

$$dC/dt = -kC \tag{6.4}$$

i.e. it is first order, where C is concentration, t is time, and k is the first order rate constant. If integrated this gives:

$$\ln[C/C_o] = -kt \tag{6.5}$$

i.e. such data should be plotted in semilogarithmic fashion if it is desired to linearize it. If plotted as

$$C/C_o = e^{-kt} \tag{6.6}$$

then a curved plot results (Figure 6.2, lower curve).

In a *suspension*, however, every time a molecule in solution decomposes, it will be replaced by one dissolving from the solid phase, so that the concentration is always C_S:

$$dC/dt = -kC_S = \text{a constant, } k_o \tag{6.7}$$

i.e., the decomposition will appear to be zero order (Blaugh and Wesolowski,1959, Grimm et al., 1973).

Assume for a given drug the solubility is 0.2 mg/mL and the first-order rate constant in solution is 2.0 year^{-1}. If a solution is tested for solubility, the data would show up, on a linear graph, as shown in Figure 6.2 (lower curve), i.e., it would be quite unstable. If a suspension of 100 mg is made, then the loss rate of the drug substance *in the suspension* will be:

$$k_o = 0.2 \times 2 = 0.4 \text{ mg/mL/year} \tag{6.7}$$

Suppose the suspension contains 100 mg of drug per mL. It will then lose 0.4 mg/mL in one year, then it will lose 0.8 mg/mL in two years, i.e., it is well above the 90% lower label claim limit.

However, if it contained only 10 mg/drug per mL. then it would still lose 0.8 mg/mL in two years, and it would then be 92% of label claim, i.e., the stability is much more marginal. This is the curve that is shown as the upper curve in Figure 6.2.

Hence we conclude two important principles: (1) stability of suspensions is zero order, i.e. *we may plot concentration retained on the y-axis and time on the x-axis and we will get a*

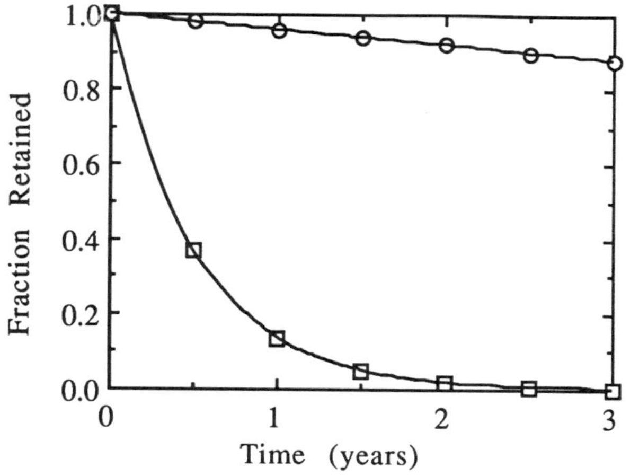

FIGURE 6.2 Data of decomposition of a solution of a compound, and a 10 mg/mL suspension of the same compound. The solution decomposes according to the equation $C = C_0 e^{-2.0t}$ and the suspension decomposes according to the equation $C/C_0 = 0.4t$, where t is measured in years.

straight line, and (2) *the stability is "better" the higher the concentration of the suspension.*

Suspensions follow Arrhenius equations, although the activation energies are the sum of the solution activation energy and the heat of solution (Carstensen, 1991).

6.6 VISCOSITY

It shall be seen in the following that the concept of viscosity is quite important. Without going into the definition of viscosity at this point, it is intuitive that it has to do with the "thickness" or "resistance to flow" a liquid has.

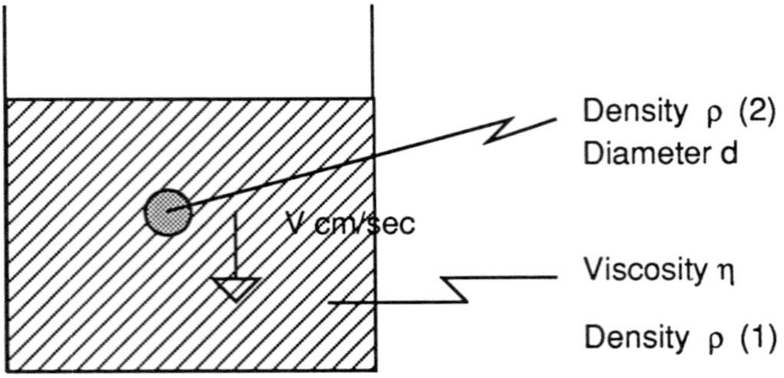

FIGURE 6.3 Stoksian fall.

One important quality is that if a sphere of density ρ_2 and diameter d is let fall in an infinitely wide and deep liquid of viscosity η and density ρ_1, then, as shown in Figure 6.3, it will, after a while, attain a constant velocity, v cm/sec, given by:

$$v = d^2(\rho_2 - \rho_1)g/(18\eta) \quad (6.8)$$

The diameter is called the Stokes diameter and the principle is used to measure viscosity in the falling ball viscometer and to determine particle size distributions in the Andreasen apparatus.

In a more sophisticated sense, one often determines so-called rheograms. If one applies a stress, σ, (pressure, force divided by area) on a suspension, and does so at a given rate, τ, then a plot of τ versus σ results in a rheogram.

6.7 SETTLING OF SUSPENSIONS

In most cases a dilute suspension will settle. There are two types of situation, one where the particles settle in a large sedimentation volume (the middle situation in Figure 6.4 and the one where they settle in a small, compact volume.

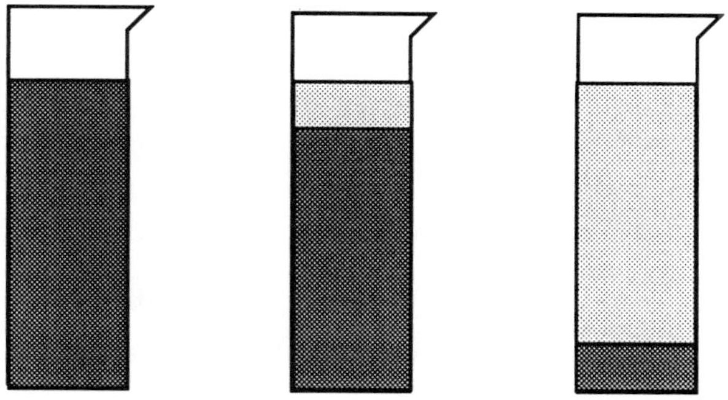

FIGURE 6.4 Types of sedimentation.

The force (F) between two particles is inversely proportional to the distance (a) between their centers squared, i.e.

$$F = Q/a^2 \tag{6.9}$$

where Q is a proportionality constant.

In the middle case in Figure 6.4, the particles are obviously further apart than in the situation to the right. The force between them is therefore smaller, so that when the suspension is shaken the particles can be evenly distributed in the liquid. In the latter case the particles are close together, the force between them is great and they are difficult to get apart, i.e., it is difficult to shake the suspension in such a fashion that it will be uniform. The latter case is referred to as caking, the former as flocculation.

It is possible to calculate the number of particles in a suspension, if all the particles are the same size (diameter d), because then their volume is

$$v = \pi d^3/6 \tag{6.10}$$

If there are X grams in the suspension and N particles, then

$$N[\pi d^3/6]\rho = X \tag{6.11}$$

i.e.

$$N = 6X/(\pi\rho d^3) \tag{6.12}$$

If, as in a caked suspension, the particles all touch one another (Figure 6.5D), then the sedimentation volume is given by Eq. 6.13.

$$V_{cake} = Nd^3 = 6X/(\pi\rho) \tag{6.13}$$

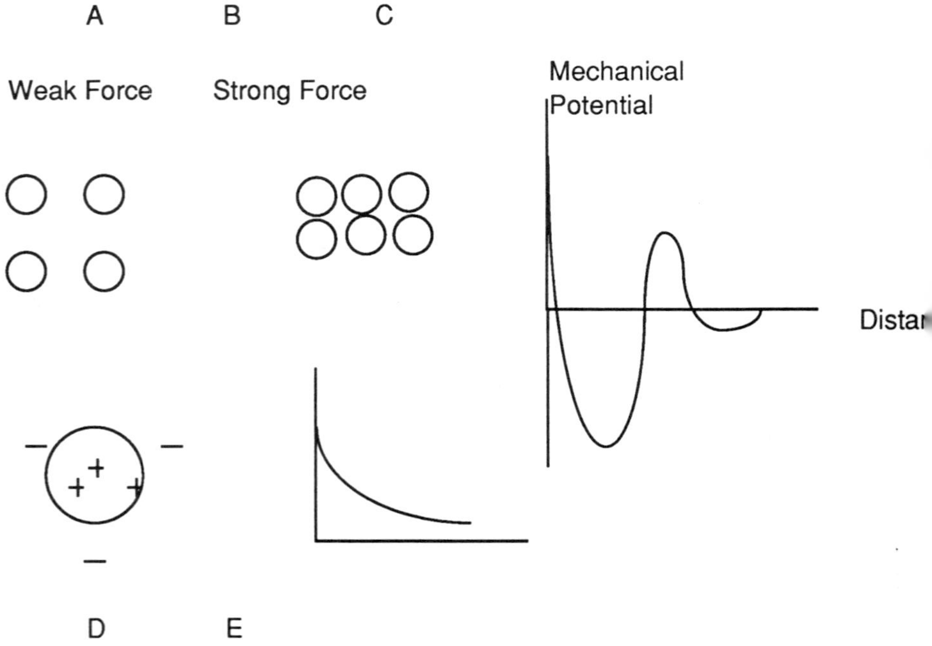

FIGURE 6.5 Potential energy of caked and flocculated suspensions.

In a flocculated suspension, the particles are *a* cm apart, and in this case the volume is given by

$$V_{floc} = Na^3 = 6Xa^3/(\pi \rho d^3) \tag{6.14}$$

so that the average distance between the particles can be calculated from Eq. 6.14.

The most desirable situation for a suspension that settles is the flocculated case. If the particles get close enough to one another at any time to "fall into" the deep potential well (Figure 6.5C), then they cannot be separated. It may take some time before they all get that close together, but it is obviously better if they could be kept in the shallow well.

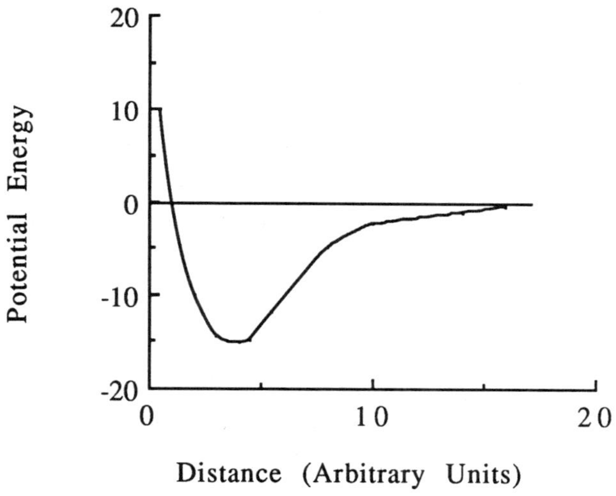

FIGURE 6.6 Potential energy curve of unadjusted suspension which is susceptible to caking.

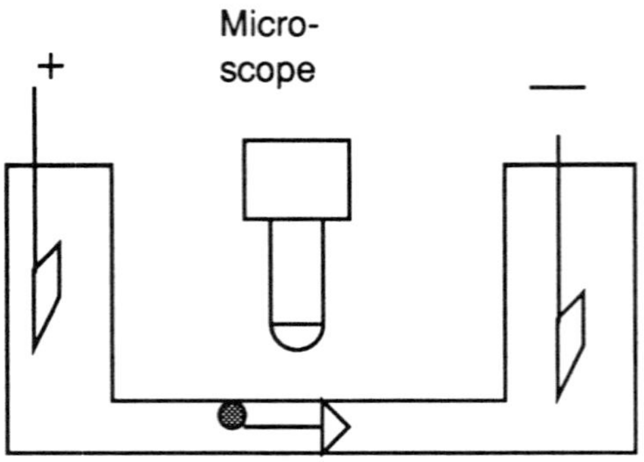

FIGURE 6.7 Zeta-potential meter.

To understand how this is accomplished, it should be noted that a suspension, of course, is electroneutral, BUT most often there is a charge difference between the bulk liquid (which is usually negative) and the suspended particles (which are usually positive). There are, of course, as many negative charges as there are positive charges (Figure 6.5D), except the positive charge is concentrated in a small volume (the solid) and the negative charges are "spread out." This is denoted an electric double layer.

In suspensions that are susceptible to caking the potential energy diagram between two particles will be as shown in Figure 6.6. There is nothing to keep two particles from getting close together. If they "start" far apart, molecular motion will (at some time) get them to get closer, the potential energy will drop, and it will keep on dropping until the particles are at the distance dictated by the minimum in the curve.

There is an electrical potential between solid and liquid (Figure 6.5D), and if two particles do get close together they will start sharing electrical double layer and hence behave like one particle. Again, this contributes to the depth of the potential energy well at low distance.

The potential energy difference is related to the so-called zeta-potential. This can be measured in a zeta-meter. (It is actually the electrophoretic potential which is measured.) In this apparatus (Figure 6.7), a potential is imposed over the liquid, and the speed with which the particles move towards one of the electrodes allows calculation of the zeta-potential. If this is adjusted close to zero, then the potential energy curve will become as shown in Figure 6.8. It is noted that there is now a "barrier" between a shallow minimum (i.e., one where the force to separate particles is small) and the deep well. The result of this is that a resting suspension will eventually show a large sedimentation volume (Figure 6.4, middle), and the particles behave as floccules, i.e., there is a certain number of particles that are aggregated into a loose assembly, which (e.g., on Stokes settling) behaves as one particle. This, of course, is larger than the individual particle, so one gets faster settling to a larger sedimentation volume with a flocculated than with a non-flocculated suspension.

6.8 SEDIMENTATION KINETICS

It is customary to check the rate with which a suspension settles. This can be done fairly easily, e.g. by measuring the interface (see Figure 6.4) as a function of time. A set of results as shown in Table 6.3 could, for instance, result.

The general state of a suspension can be deducted from data of this type. First of all they are plotted as shown in Figure 6.9.

First of all, the curve levels off at some unknown (at this point in time) value. This value is important, because this is the sedimentation value. With the number of data points in Figure 6.9 it is not difficult to estimate that the curve will level off at 5 cm. But if, for instance, only data after 6 months storage were at hand, then such an extrapolation would be difficult (Figure 6.10).

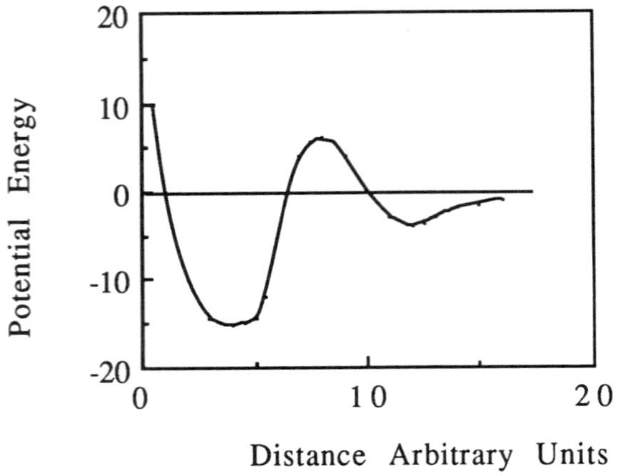

FIGURE 6.8 Potential energy curve after zeta-potential adjustment.

In such a case, it is important to know the functional relationship (Carstensen and Su, 1970a, 1970b; Michaels and Bolger, 1962) for such sedimentation curves. It is

$$H = H_{\infty} + Ae^{-at} + B^{-bt} \quad (6.15)$$

TABLE 6.3. Sedimentation Height as a Function Of Time.

Time mo	Height cm	Time mo	Height cm	Time mo	Height cm
0	30	6	15.010	12	9.370
1	26.330	7	13.660	14	8.380
2	23.240	8	12.520	16	7.630
3	20.640	9	11.540	18	7.060
4	18.440	10	10.710	20	6.630
5	16.580				

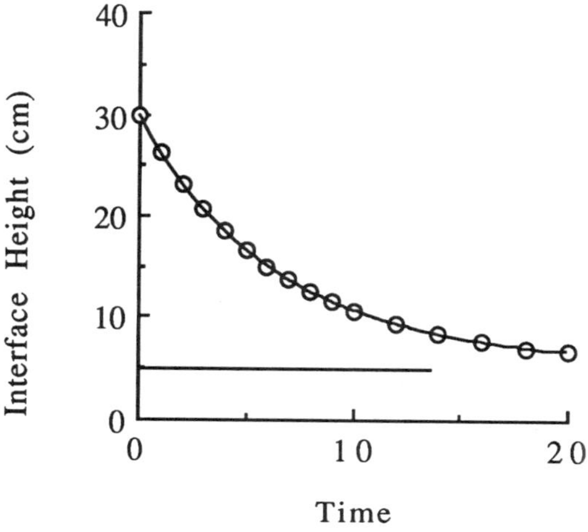

FIGURE 6.9 Sedimentation height as a function of time in months.

It is necessary to use a non-linear fitting program to obtain the values of the four parameters, A, a, b, and H_∞. It is noted that there are not five, because there is the restriction that

$$H_0 = H_\infty + A + B$$

where H_0 is the initial height and $H = H_\infty$ is the final height (corresponding to the sedimentation volume).

In this manner it is possible, fairly early in the development, to assess if one has a suspension that potentially could cake. *It should be pointed out that suspensions of caking propensity often give content uniformity problems, even at an early stage.*

It can be shown for the above system that the equation is

$$H = 5 + 10e^{-0.1t} + 15e^{-0.2t} \tag{6.16}$$

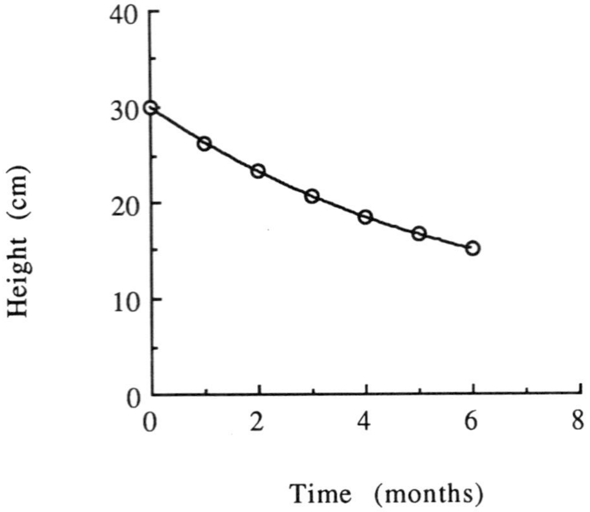

FIGURE 6.10 Sedimentation height as a function of shorter times.

6.9 DENSITY MATCHING

It is obvious from Stokes' law that if the liquid and the solid have the same density, then settling does not occur. There have been attempts to, in this manner, make suspensions, however, the principle doesn't work well, because the density of a solid is only slightly affected by temperature, whereas the density of a liquid is greatly affected, so that if the density matches at, e.g. 25°C, it will not match at 25.2°C, and so settling will still occur.

6.10 PARTICLE GROWTH

Completely isothermal storage is never possible, and the solubility changes with temperature, mostly in an increasing fashion.

Small particles have the larger specific surface, so they will preferentially dissolve when the temperature increases, but precipitation occurs primarily on the larger particles, so that, on standing, there is the potential for particle growth. This is referred to as "ripening" or Ostwald ripening.

Dissolution slows down with increasing viscosity, as does precipitation rates. Water-soluble, high molecular weight compounds (gums) increase viscosity, and hence protect the system from temperature induced particle growth.

Particle size distributions are usually obtained by way of electronic counting methods (Coulter, Malvern), but the problem is that the sample must be strongly diluted before the counting can take place.

6.11 RHEOGRAMS

Eq. 6.8 is repeated at this point for convenience and the dimensions of each term is listed in Eq. 6.18.

$$v = d^2(\rho_2-\rho_1)g/(18\eta) \qquad (6.8)$$

or

$$\eta = d^2(\rho_2-\rho_1)g/(18v) \qquad (6.17)$$

$$\mathrm{Dim}(\eta) = \mathrm{cm}^2\ (\mathrm{Mass/cm}^3)\ \mathrm{Acceleration}/(\mathrm{cm\ sec}^{-1}) \qquad (6.18)$$

$$= \mathrm{Force/cm}^2 \text{ divided by } \mathrm{sec}^{-1} \qquad (6.19)$$

Instruments that are able to measure the rate with which a force is applied to the suspension as a function of the stress (force per area) applied to it should then give the viscosity as the slope of the curve. Curves that are straight lines through the origin are denoted Newtonian, if they have curvature to them, then they are pseudo-plastic (Figure 6.11). A Bingham body gives a fairly straight line, but does not intercept at the origin, and the intercept with the x-axis is denoted the yield value. Below this stress, there is no movement of the mass.

In a rheogram, the stress in cgs units (dyn/cm^2) or IU's (Pascals) is plotted as the abscissa and the shear rate in reciprocal seconds is plotted as the ordinate. An actual viscosity (cps in CGS system) can be obtained as the slope of the curve at the particular stress. *Note that with a Brookfield viscometer, a scale reading is obtained, but this is not convertible to absolute units*, and it is difficult to obtain even a relative rheogram when spindles are changed.

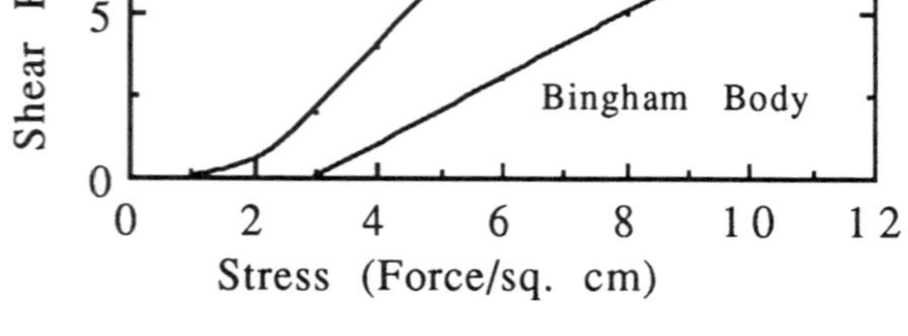

FIGURE 6.11 Pseudoplastic and Bingham body behavior.

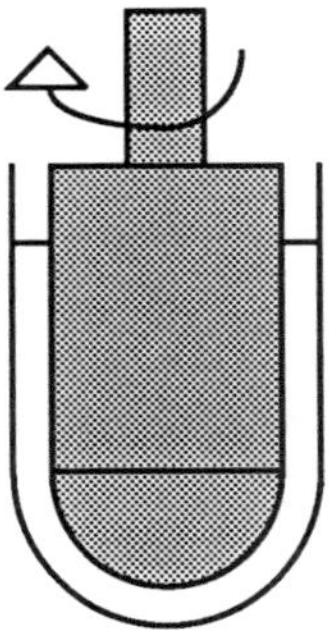

FIGURE 6.12 Cup and Bob Method.

It shall be seen shortly that it is exceedingly important to obtain yield values for suspension systems, and a cup and bob method (Figure 6.12) is advocated for all suspension development work. For routine release testing (post NDA), one or two point relative methods such as a Brookfield are adequate.

For control of raw materials used in suspension, e.g. protective colloids, full rheograms are advised, since gums are natural products and subject to seasonal variation.

6.12 YIELD DIAMETER OF SUSPENSIONS

As seen above, the yield value of a suspension is a function of the liquid in which the solid is suspended, but also of the solids fraction in the suspension. The higher the solids fraction, the higher the yield value.

The yield value in a suspension is important, because if the sum of forces acting on the particle is less than the yield value

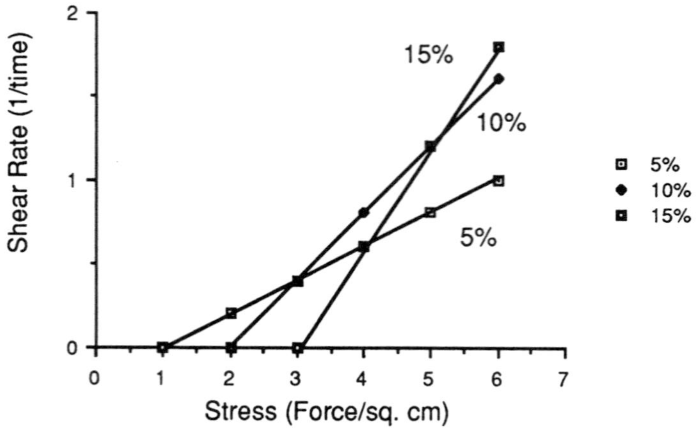

FIGURE 6.13 Rheograms of suspensions of different solids fraction.

times its cross-sectional area (the force acting on it), then it will not move, so, *in that case, there is no settling.*

The forces acting on the particle are: gravitational; forces between the particles; and forces between the particles and the wall of the container that houses them. This latter is larger the "closer" the particles are to the wall, and if the diameter is small, then this force is large.

The smaller the diameter of the tube that houses a suspension, the higher will be the sedimentation volume if the suspension is (even partially) flocculated. This is shown in Figure 6.14.

The point of interest is at or below which diameter there is no settling. This is denoted the yield diameter, d*, and is given by:

$$d^* = 4\tau/g\Delta\rho)\phi \tag{6.20}$$

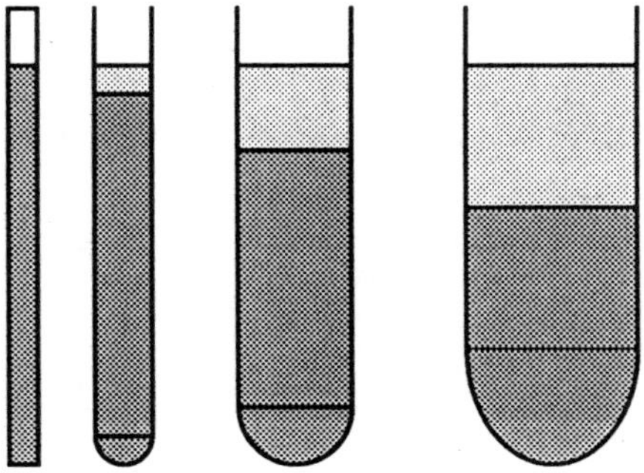

FIGURE 6.14 Sedimentation volumes as a function of tube diameter.

where τ is the yield value and ϕ is the solids fraction. It is noted that there is a strong effect of concentration, because an increase in ϕ by itself makes d* larger, but it also increases τ, so that the concentration in two ways causes an increase in yield diameter.

This shows the importance of bottle size and the danger of changing concentrations of solid matter in a suspension.

In a practical sense one may make a series of long tubes, such as in the figure above, and place a suspension in it, and then extrapolate the interfacial height to where it is equal to the original height of the suspension. This is shown in Figure 6.15.

It is seen in Figure 6.15 that it would be desirable to extrapolate the curve to see at what diameter (the yield diameter) H/H_0 is 1.0. This is not that easy with data that exhibit curvature. Sometimes one may try to assign a function to the curve. It is obvious that there is a minimum sedimentation volume (at high diameter), and above this diameter it will stay virtually constant. If this is taken to be 0.4, and is assumed to be logarithmic, then the curve in Figure 6.16 results.

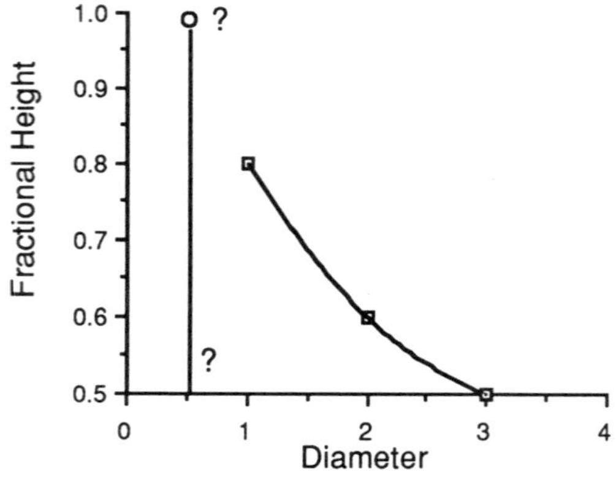

FIGURE 6.15 Height fraction (height of interface/initial height) as a function of tube diameter.

It is noted that the extrapolated diameter is negative, i.e. the suspension in question does NOT have a yield diameter. It is not all suspensions that have yield values.

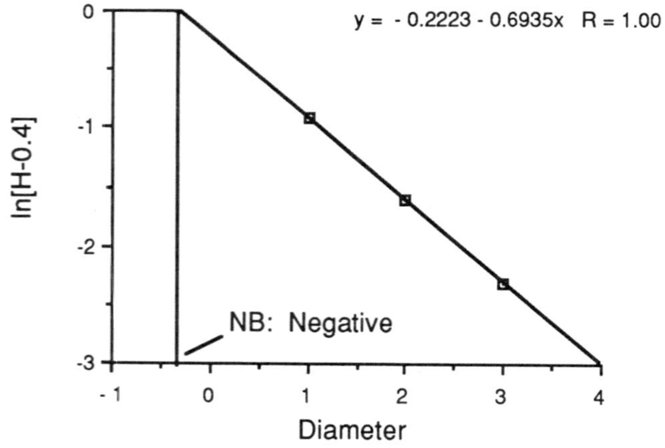

FIGURE 6.16 Extrapolation of the last three heights in Figure 6.15.

The functionality assigned is:

$$\ln[(H/H_o)-0.4] = -q(d-d^*) \qquad (6.21)$$

Of course this can be used for non-linear (iterative) programming. The functionality may, however, not be correct (even though the fit is good), and is only presented here as a suggestion.

6.13 PROTECTIVE COLLOIDS

Colloids, gums, are often employed in suspensions, and these serve several functions.

(a) They attach to the surface of the solid, and hence, effectively, increase the diameter, so that settling rates are decreased.

(b) They also exist as non-adsorbed polymers in the solution, and as such they increase the viscosity and yield value of the suspension.

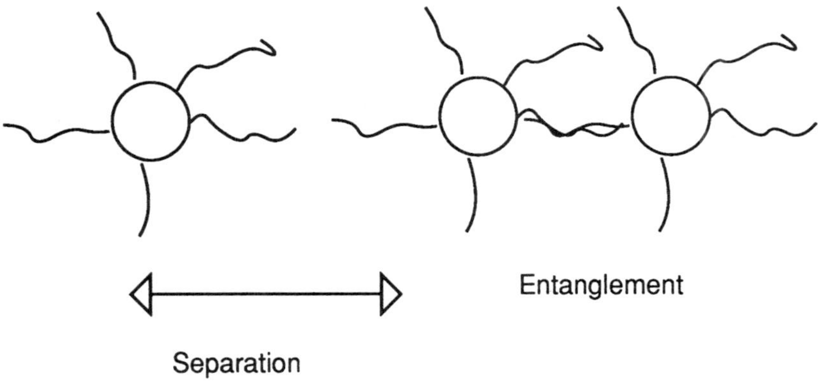

FIGURE 6.17 Colloids that separate and that entangle.

(c) They "keep the solids particles apart." If the gum is properly chosen, the particles can come no closer than two times the length of the polymer chain. Improperly chosen they can, however, cause entanglement which is a secondary cause for caking.

6.14 ADSORPTION ISOTHERMS

Solutes will adsorb onto the solid phase, and this is of importance. For instance, the preservative (parabens for instance) are effective in the solution in the concentration in which they are there. However, there is part adsorbed, so that the concentration in solution will be less than the added amount divided by the liquid volume.

It is prudent to conduct an adsorption experiment. Here a given amount of solid is in suspension and the total amount of solute is dissolved in the liquid before the solid is added. This is equilibrated, centrifuged (not filtered, since material may be adsorbed on the filter), and the concentration in solution is determined (column 2 in Table 6.4). Knowing the total amount added, and the amount remaining in solution, the amount adsorbed is then calculated by difference.

TABLE 6.4 Results of an Adsorption Experiment

C_{total}	$C_{solution}$	C_{solid}	$\ln[C_{solution}]$	$\ln[C_{solid}]$
15.975	11.000	4.975	2.398	1.604
32.5	25.000	7.500	3.219	2.015
38.216	30.000	8.216	3.401	2.106
51.721	42.000	9.721	3.738	2.274
66.124	55.000	11.124	4.007	2.409

Such data, in solution, most often follow a Freundlich isotherm:

$$C(solid) = n[C(solution)]^{1/p} \tag{6.22}$$

where C(solid) is the concentration of adsorbed material (g/g) and where C(solution) is the concentration of solute. This may be written

$$\ln[C(solid)] = \ln[n] + (1/p)\ln[C(solution)] \tag{6.23}$$

Each solute (e.g., preservative) has its own characteristic value of p and n. If the data above, for instance, applied to sodium benzoate, then they may be plotted as shown in Figure 6.18. If it is desired to have a concentration of 10 mg/mL, then one may calculate:

$$\ln[C(solid)] = 0.40547 + (0.5\ln[10]) = 1.5568 \tag{6.24}$$

so

$$C(solid) = \exp(1.5568) = 4.76 \tag{6.25}$$

i.e. a total of 14.76 mg/mL would have to be added to allow a concentration of 10 mg/mL in the solution phase.

The calculation is only strictly correct if the volume of solid is small. If it is larger, then the volume of liquid is smaller and this must be taken into account. In that case the total amount would be smaller than if the adjustment had not been made.

At times the adsorption from solution will follow a Langmuir isotherm, given by the equation

$$1/C(solid) = A + [B/C(solution] \tag{6.26}$$

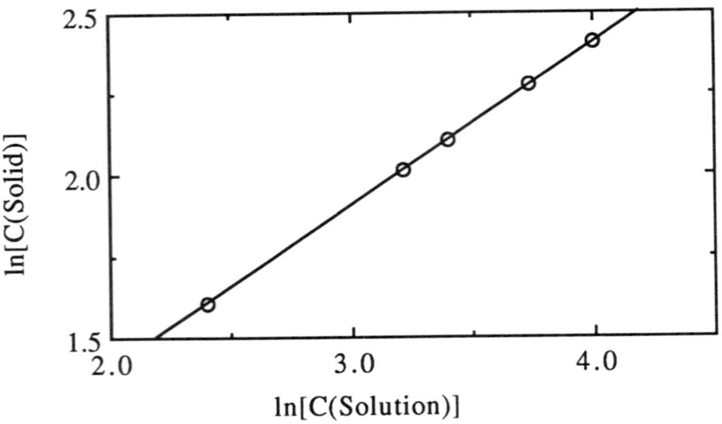

FIGURE 6.18 Freundlich adsorption isotherm.

It can be difficult to distinguish the two types. For instance the data in Table 6.4 has been treated by Eq. 6.26 in Figure 6.19, and it is seen that the fit is quite good.

6.15 COASCERVATES

Many colloids have an isoelectric point. This means that they have a positive charge and a negative charge in the molecule. It often happens with proteins, where a COOH group will give up an H^+ and change to COO^- whereas the H^+ attaches to an NH_2 group and becomes NH_3^+.

If two protective colloids are used then there is always the danger of coascervation. At a given pH, A may be below its isoelectric point and B above, so that A is primarily positively charged and B negatively charged. A compound A^+B^- may then form, and this may either have higher viscosity or actually precipitate, forming strings in the suspension.

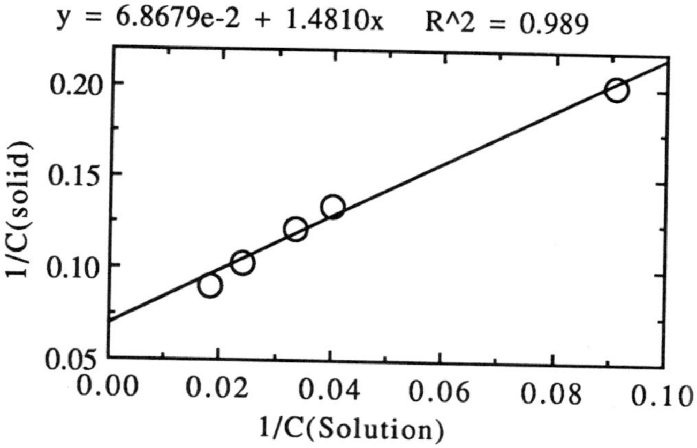

FIGURE 6.19 Example of Langmuir isotherm. Data from Table 6.4 treated by Eq. 6.26.

In general, protective colloids are subject to physical change with pH, and it is wise to test the two protective colloids at the intended pH of the suspension, by themselves, before choosing them in the system.

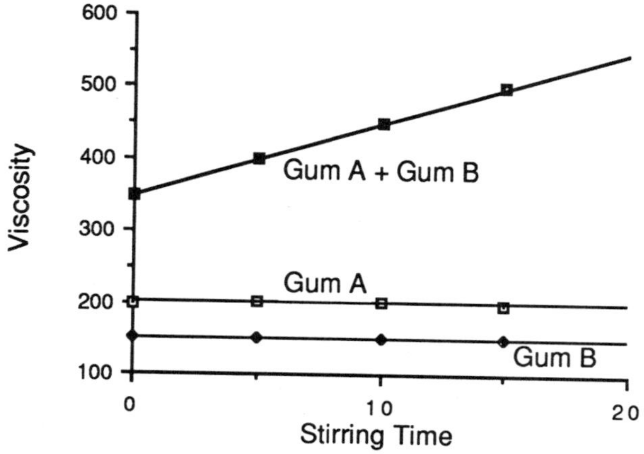

FIGURE 6.20 Testing for colloid interaction.

Figure 6.20 shows how viscosity changes as a function of length of stirring when two colloids are present at a given pH versus the lack of change for the individual colloids. The viscosities here are at a given shear rate.

6.16 SURFACTANTS

Surfactants are used in suspensions for several reasons:

(a) They make wetting of the solid easy during manufacture.
(b) They also add in particle separation.
(d) They may or may not improve dissolution rates.
(e) They add to aeration propensity, which is a disadvantage.
(f) They may interfere with preservative activity.
(g) They may cause solubilization of the drug, which, in general, is a disadvantage.

Particularly points (e) and (g) are important, and if one changes surfactant or surfactant concentration, these properties should be re-investigated.

6.17 PRESERVATION AND SURFACE ADSORPTION

There are several solutes in the liquid phase of a suspension. Each of these will adsorb unto the solid, and if each of them were present by themselves, they would follow either a Freundlich or a Langmuir isotherm, and amounts on the solid and in solution could be calculated.

For preformulation purposes it is of importance to determine the distribution of the preservatives (e.g., parabens) between liquid and solid phase with "reasonable" levels of protective colloids and surfactants present.

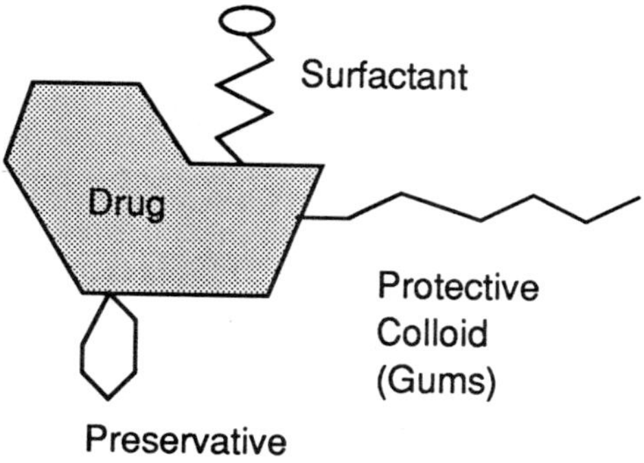

FIGURE 6.20 Adsorption types on a solid in suspension.

When a final formula is arrived at, such a study should be repeated. It should be recalled that the effective concentration of preservative must be present in solution (not in toto).

6.18 DISSOLUTION OF SUSPENSIONS

Dissolution testing will be covered in more detail at a later point in this text, and the reader, if not familiar with dissolution may want to refer to these later section prior to reading the following.

Dissolution testing of suspensions has been required by the FDA for the last decade. If a suspension is allowed to flow into a dissolution apparatus, two things may happen as shown in Figure 6.22. It may disperse, as indicated in the figure below at the situation below the arrow, or it may not disperse. In the former case a cube root dissolution pattern results:

$$1-(M/M_o)^{1/3} = 1-(F)^{1/3} = Kt \qquad (6.27)$$

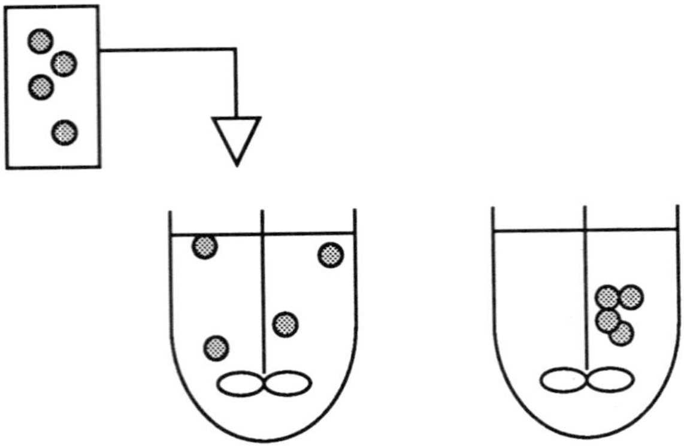

FIGURE 6.22 Dispersion modes in a dissolution apparatus.

If percent is used, then

$$4.634 - F^{1/3} = K^{*}t \qquad (6.28)$$

where F is the fraction not dissolved. Data of this type are shown in Table 6.5 and Figure 6.23.

If, however, the material agglomerates, then it will dissolve by a square-root in time law (the Higuchi equation):

$$Q = L\sqrt{t} \qquad (6.29)$$

If the data in the table are treated in this fashion, the fit is also quite good. Finally, many data sets follow (Cárdenas et al., 1994):

$$\ln[100M/M_o] = \ln[100] - K'(t-t_i) \qquad (6.30)$$

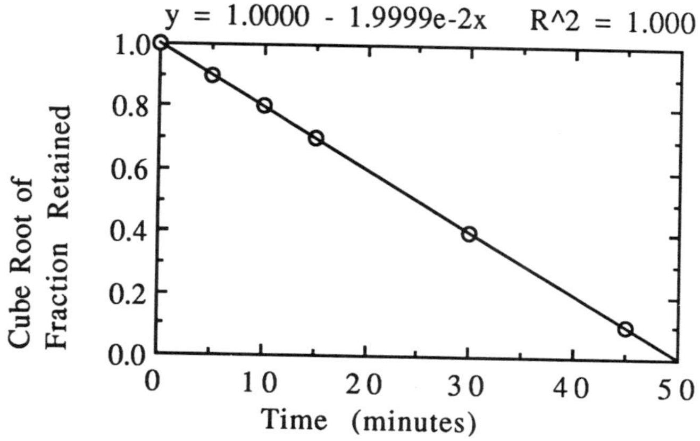

FIGURE 6.23 Cube root dissolution data of a suspension.

If the data from the table are treated by either of these, a reasonable fit will result, so that mere curve-fitting will not tell which of the situations occur. This is shown in Figure 6.24.

The last form of the equation is the preferred one from biopharmaceutical points of view, because it allows direct correlation with curve-fit blood level data.

Table 6.5 Dissolution Data Treated Via a Cube Root Equation

Time (min)	Percent Not Dissolved F	$F^{1/3}$
0	100	4.634
5	72.9	4.171
10	51.2	3.708
15	34.3	3.245
30	6.4	1.855
45	0.1	0.465

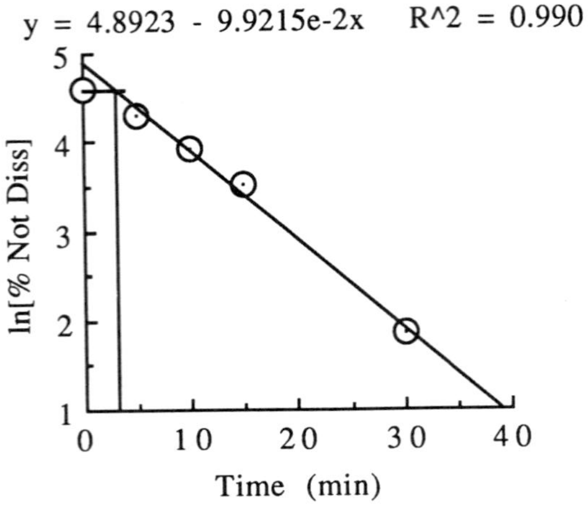

FIGURE 6.24 Data from Table 6.5 plotted according to Eq. 6.29.

6.19 CONTENT UNIFORMITY

It is advantageous to set up a mechanical method of "shaking," rather than simply shake and see what happens. Such a test could be placing the suspension bottle on a shaker, or on a rotating shaft, and then exert "shaking" for a given length of time. It should be noted that such shaking methods should be validated from time to time.

To test for resuspendability, a sample is taken from one bottle at the top, from another from the middle and from a third from the bottom.

Visual observation (cake or not?) should also be done at the same time.

6.20 PRE-NDA AND PRE-ANDA TESTING

Some or all of the above principles are applied during the formulation and preformulation stages of product development, for example the determination of the adsorption isotherm of preservative on the suspended drug substance. But many of these are attributes that add to the knowledge base about the product, that have been used as an aid in arriving at the final formula. One test that is often carried out in the formulation stage is 5 time point dissolution, but once this is established, a 2 point dissolution should suffice for stability testing pre-NDA. So for routine pre-NDA testing, many of the formulation "tests" are not necessary.

The following testing is recommended prior to the NDA or ANDA. The time periods at which they are carried out may vary depending on the product. In the strictest case, all tests could be carried out at the usual 3, 6, 9, 12, 18, 24, 36 month pull-times, but aside from assay, dissolution, viscosity and taste, most of the tests could be restricted to a yearly basis.

Bacteriology (including challenge tests)
Assays of drug substance and preservative
Content uniformity as a function of time
One point viscosity as a function of time (Brookfield, Ostwald-Fenske)
Two point dissolution (about the time where half is dissolved and at 30 min) as a function of storage time
Density
Microphotography
Surface tension
LD50 (only occasionally)
Particle size
Appearance
Taste
Cap torque
Weight

Of these, the last two are really packaging tests, but they are easily carried out, once the product has been pulled. Samples for weight should have noted on them the initial weight.

If the suspended matter disappears or greatly diminishes at 40°C either initially or as time passes by, this should be pointed out in the NDA and used as a justification for not including it in further routine testing. The 40°C test, per se, is not very meaningful, but it is particularly not so if the physical state of the product is altered at the accelerated station. Of course, the FDA has no obligation to accept the recommendation.

Several properties are bound to change at 40°C testing, for instance particle size distribution. An alternative to the six month test at 40°C is, of course, one year at 30°C, and in general this may be a much better alternative, since many well-formulated suspensions would not stand the testing at 40°C.

6.21 INITIAL SPECIFICATIONS

The following is a list of tests that could be suggested for the release of manufactured batches. Proper statistical limits should be set on each (and justified).

Bacteriology (including challenge tests)
Assays of drug substance and preservative
Content uniformity
One point viscosity as a function of time (Brookfield, Ostwald-Fenske)
One point dissolution (Q_{30} for instance)
Density
Microphotography
Particle size

Appearance
Taste
Cap torque
Weight

6.22 TESTING POST-NDA AND POST-ANDA

Less testing should be suggested in the routine protocol:

Bacteriology (yearly) or assay of preservative
Assay of drug substance
Content uniformity (yearly at most)
One point viscosity as a function of time (Brookfield, Ostwald-Fenske) (yearly)
One point dissolution (Q_{30}) as a function of storage time
Appearance (yearly)
Taste (yearly)
Cap torque (yearly)
Weight (yearly)

It is noted that particle size is recommended for initial testing but not for post-NDA testing, because in most cases it is, by thermodynamical principles, bound to change and hence does not reflect on the stability of the product.

This, of course, does not prevent the FDA from requiring that it be done.

6.23 PRODUCTION PROBLEMS

A general flowchart is shown in Figure 6.25. (Sweeteners, flavors, and dyes are not shown). Specific problems can be specks of dirt and air inclusion. If the size of the specs is larger than the individual suspended drug particle, it can be removed by passing the suspension through a rotating stainless steel filter

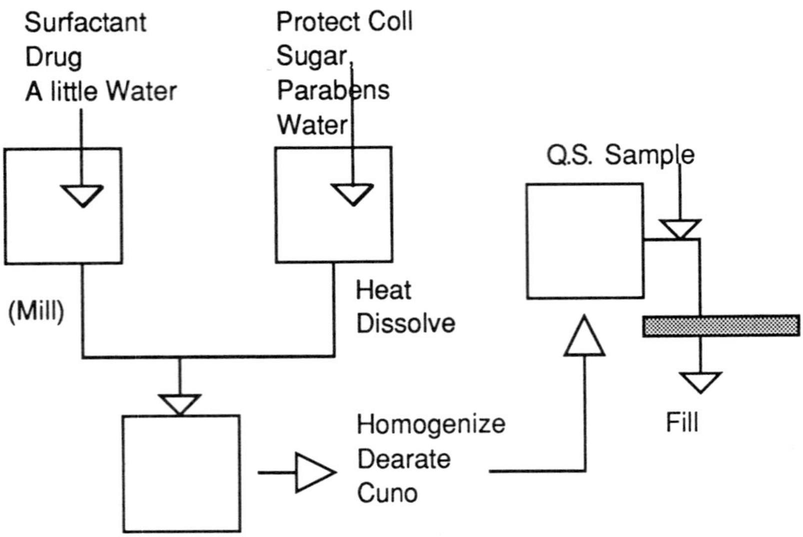

FIGURE 6.25 Flowchart of suspension manufacture.

(Cuno). Occluded air can be drawn out, for instance, by vacuum in a rotating pan (Versator).

The q.s. step should be done after deaeration and is often done by weight. Air inclusion may still occur in pipes leading from the holding tank to the filling line.

Although air is not checked for in the usual specification for release, excessive air will result in too low a density.

Liquid mills, it should be emphasized, rarely do size reduction, but they deagglomerate the drug (e.g., in the transfer in the first flowchart step).

REFERENCES

Blaugh S.M. and Wesolowski, J.W., (1959), *J. Am. Pharm. Assoc., Sci. Ed.*, **48**: 691.

Cárdenas, R.H., Cortés, A.S., Artogge, R.R., Luna, M.P. and Dominguez, R.A., (1994), *Drug Dev. Ind. Pharm.*, **20**:1063.

Carstensen, J.T., (1995), *Drug Stability,* 2nd Edition, Marcel Dekker, New York, pp.187-188.

Carstensen, J.T. and Su, K., (1970a), *J. Pharm. Sci.*, **59**: 666.

Carstensen, J.T. and Su, K. (1970b), *J. Pharm. Sci.*, **59**:670.

Grimm, W., Hugger, J.P. and Neugebauer, I., (1973), *Pharm. Ind.*, **35**: 733.

Michaels, A.S. and Bolger, J.C., (1962), *Ind. Eng. Chem. Fund.*, **1**:24.

CHAPTER 7

Crystalline Solids

Solids are either amorphous or crystalline. Chapter 7 deals with crystalline solids. Amorphates will be treated elsewhere.

Crystalline substances are characterized by existing in crystal lattices. These are defined as periodic arrays (Figure 7.1). They are defined by lattice constants, a, b, and c, which may be identical or different, and by angles, which may or may not be equal to 90°.

7.1 CRYSTAL SYSTEMS

To get an idea of what the order of magnitude of a lattice constant, a, would be, consider for instance aspirin. Aspirin's molecular weight is 180 and its true density is 1.25 g/cm^3; therefore its molar volume is 180/1.25 = 144 cm^3/mole. There are 6×10^{23} molecules in one mole of a substance, so that it is possible to calculate the volume of one aspirin molecule to be $144/(6 \times 10^{23}) = 240\ 10^{-24}\ cm^3 = 240\ Å^3$. If it is assumed that

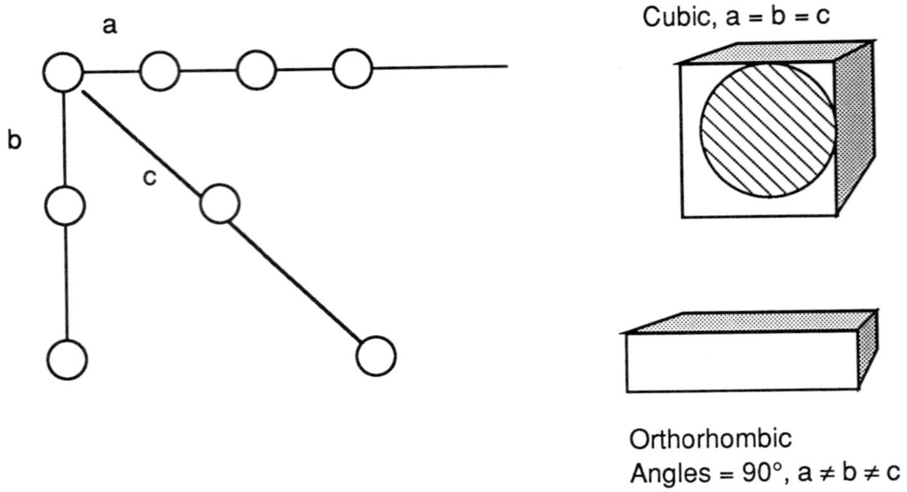

FIGURE 7.1 Example of crystal lattices.

the molecule (Figure 7.1) is "encased" in a cube, then distance a is $[240]^{1/3}$ or 6.2 Å. This then would be the molecular spacing. In actuality the distances between layers of molecules in a crystal may be obtained from X-ray diffraction (Figure 7.2).

The principle of such a measurement is obtained from Bragg's law: Two rays with a given wavelength, λ, are reflected from the surface and from the first molecular layer of a crystal; i.e., the top one is reflected directly from the surface and the second one from the first layer of molecules.

There will be waves traversing the first layer and passing on to the second layer and then reflected, and these will travel a distance of 4[asin(ø)] longer than the first. If they are in phase (i.e., their maxima are at λ distance from each other), they will reinforce one another and their intensity will double, so that when the angle is correct there will be a peak at the collimator. For waves that travel to the second layer and are reflected, the

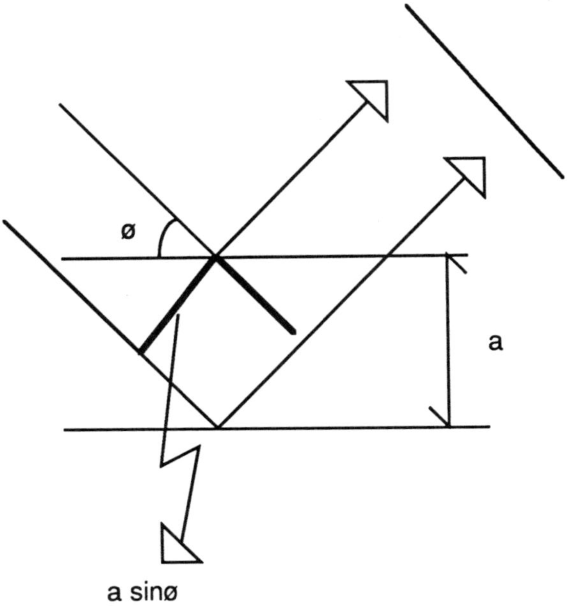

FIGURE 7.2 X-ray impacting planes of molecules.

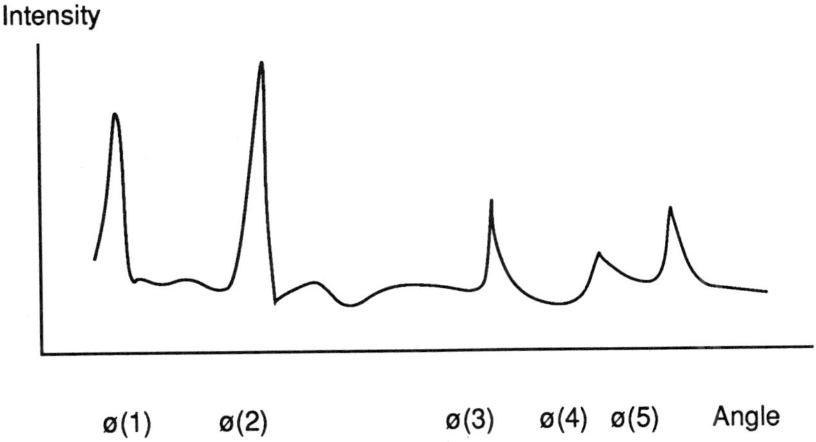

FIGURE 7.3 X-ray diffraction pattern.

amplification will be less, but it follows in general that the angles at which there are amplifications are given by:

$$2\ [a\sin(\phi)] = n\lambda \qquad (7.1)$$

where n is an integer.

In the X-ray experiment the wavelength is fixed and the angle is changed. One notes (Figure 7.3) the angles of amplification ($\phi(1)....\phi(5)$) and now can calculate a if n is known.

One can show that if n = 1, then the angle is quite shallow, and penetration (which is the smaller the smaller the angle) is marginal, so that the diffraction patterns one obtains are "harmonics," i.e., correspond to multiples of the spacing.

In pharmaceutics this is of no importance, because X-rays are used to differentiate between different crystal systems and different types of crystals and crystal spacing, so that comparison between two diffraction patters will ascertain whether or not two different so-called *polymorphs* are at hand or not. The phenomenon of polymorphism will be discussed a bit further down in the chapter text.

7.2 DRUG PURIFICATION

When a drug is synthesized it contains impurities (intermediates, reactants), and these are removed in the final step (the polish), which is a recrystallization.

It should be pointed out that an *impurity* in general is different from a *decomposition* product. These latter are produced during the storage of the drug and the drug dosage form, although they are at times the same substance, e.g., salicylic acid occurs as an impurity in aspirin and also as its main decomposition product. Aspirin anhydride, on the other hand, is an impurity because it is not formed on storage at room temperature.

TABLE 7.1. Solubility of Sulfanilamide in Water.

Temp (°C)	Solubility, g/1000 g water
10	2.6
25	7.5
60	40

Purification takes place, primarily, by two processes: recrystallization or reprecipitation. This is illustrated below. Sulfanilamide exhibits the solubility shown in Table 7.1 at different temperatures (Merck Index).

Assume that 80 g of sulfanilamide are dissolved in 2000 g of water at 60°C and that the solution is then filtered if not all the solid goes into solution. The solution is then let cool to 25°C, in which case the amount of sulfanilamide that can stay in solution would be 15 g, so that theoretically 80 - 15 = 65 g of sulfanilamide would precipitate out.

Most of the time the solubility of the impurities (assumed present in low concentration) would not be superseded, i.e. the sulfanilamide crystals that precipitated out would be fairly pure. Upon slow cooling the crystals will be fairly large; on fast cooling they will be fine.

It might seem desirable to make the crystals small, because this would result in a large specific surface area. This, in turn, would give rise to rapid dissolution which should be advantageous. There are, however, adverse effects of fine particles as well, poor flow, for instance, and for tablets this will result in capping tendencies. A controlled particle size (being optimum for most properties) is usually what is strived for.

Purity of drug substances is determined e.g. (a) by chemical assay (for instance HPLC) (b) by phase solubility (c) by differential scanning calorimetry. These give results that are quantitative, whereas (d) thin layer chromatography, gives but qualitative results.

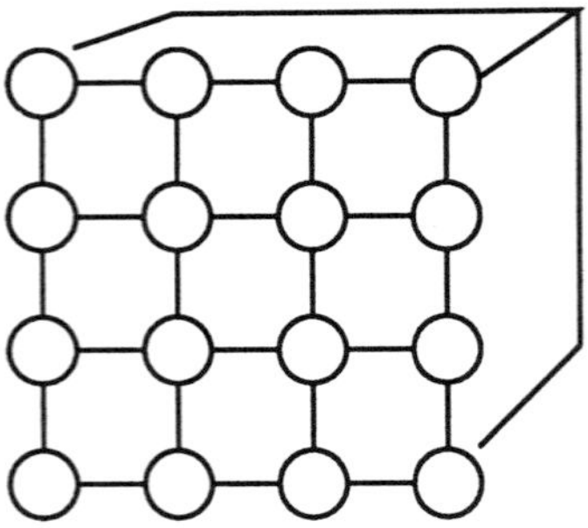

FIGURE 7.4 Cubic lattice. There is a molecule in each shown corner.

Precipitation is a purification method which may be best described by example. At 10°C one gram of sulfanilamide dissolves in (i.e. is soluble to the extent of) 5.1 cm^3 of acetone. If 94.9 cm^3 are added, then the 100 cm^3 will be "nearly pure" water, i.e., the solubility would be about 2.61 g/liter, i.e. 0.261 g in the about 100 cm^3 of solution. Therefore, 1-0.261 = 0.739 gram of a more pure material will precipitate out. Drug substances that are purified in this way are almost always of a very small particle size.

There is a total of seven crystal systems. It is not necessary for purposes here, to know their names, but rather to know of their existence and how they differ. The cubic system (Fig. 7.4), is one where all sides equal one another and where all angles are 90°. In the orthorhombic system all angles are 90° but the side lengths are unequal. The monoclinic system is one where all sides are different, where two of the angles are 90° but where one is not. The other systems combine non-right angles and unequal or equal side lengths; they are: triclinic, tetragonal, trigonal and hexagonal.

An inorganic compound usually crystallizes in a specific crystal system. Sodium and potassium chloride exist in the cubic system. Sulfur is an exception in that it can be both monoclinic and rhombic. Conversely, organic compounds invariably can crystallize in several different crystal systems. The governing factor in which solvent is employed in the recrystallization or what solvent pair is employed for reprecipitation. This phenomenon is called polymorphism. This is important pharmaceutically and will be treated shortly in the discussion further in the text.

7.3 CRYSTAL HABIT

When viewed under the microscope a crystal will exhibit different "faces." When viewed macroscopically, different shapes will occur, but these do not necessarily indicate that they are different crystal systems.

When precipitation from a supersaturated solution takes place, the first step is the occurrence of a nucleus. This then grows (Figure 7.5), and if the growth is equal in all directions, then a cubical shape will result. When the growth is slower in one direction, a plate will result, and if it is slow in two directions, a needle shape results. Shape differences do not signify different crystal systems, but rather are denoted crystal habits. They are of importance because a cubic-like shaped multiparticulate powder flows more readily in a powder hopper. This is of importance in tableting. Conversely a less "cubic" crystal will have a larger specific surface area, and in turn the dissolution rate is favorably affected.

7.4 VAPOR PRESSURE OF SOLIDS

Most solids have low vapor pressures. Two solids, of pharmaceutical importance, nitroglycerin and ibuprofen, have vapor pressures that are easily measurable at room temperature.

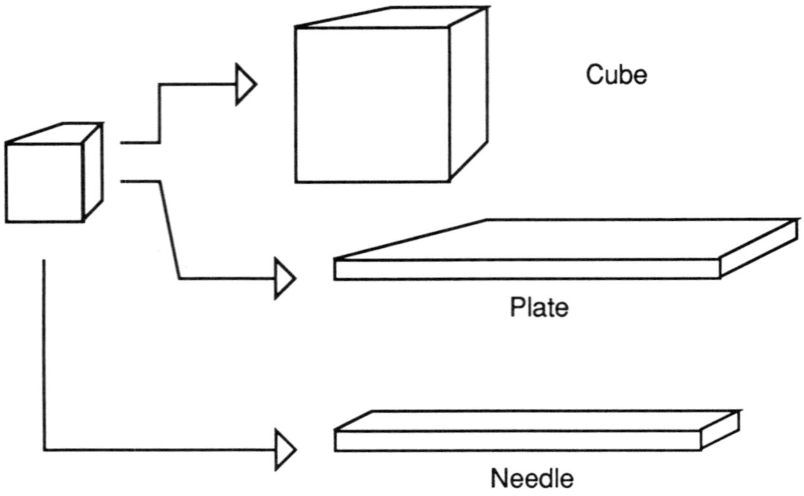

FIGURE 7.5 Nucleus growth.

The vapor pressures, P, of a solid, will increase with absolute temperature, T, as described in the Clausius-Clapeyron equation:

$$\ln[P] = -\{\Delta H^*/R\}(1/T) + \phi \qquad (7.2)$$

Here ΔH^* is the heat of sublimation, ϕ is a constant and R is the gas constant. Vapor pressures of benzoic acid at different temperatures are listed in Table 7.2. The data are plotted in Figures 7.6 and 7.7.

Henry's law (Eq. 7.3) is applicable in the situation shown in Figure 7.8. A compound is present in the gas phase and has a vapor pressure, P'. It will partition between the gas phase and the liquid, where its (molal) concentration is C. Henry's law expresses the concept that these two quantities are proportional, i.e., that:

$$C = KP' \qquad (7.3)$$

TABLE 7.2 Vapor Pressure Data for Benzoic Acid

Temp (°C)	1000/T	P (mm Hg)	lnP
25	3.35	0.0071	-4.95
45	3.14	0.036	-3.32
65	2.96	0.15	-1.90
85	2.79	0.53	-0.63

This can be rewritten

$$\ln C = \ln[K] + \ln[P'] \quad (7.4)$$

When solid is present, then the solution must be saturated. The vapor pressure of the solid will be equal to its natural vapor pressure, P, indicating that:

$$\ln[S] = \ln[K] + \ln[P] \quad (7.5)$$

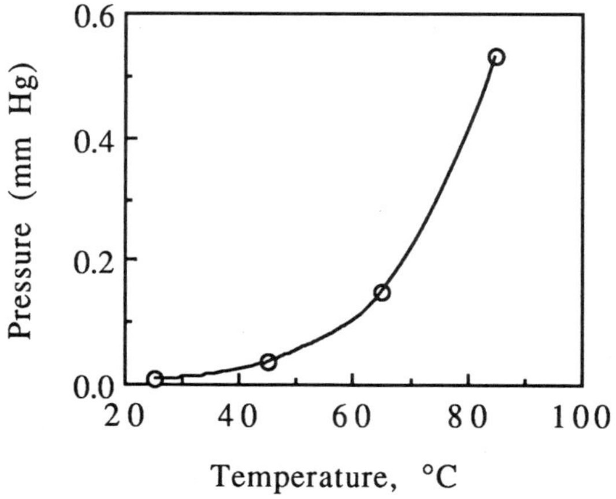

FIGURE 7.6 Vapor pressure of benzoic acid as a function of temperature. See Table 7.2.

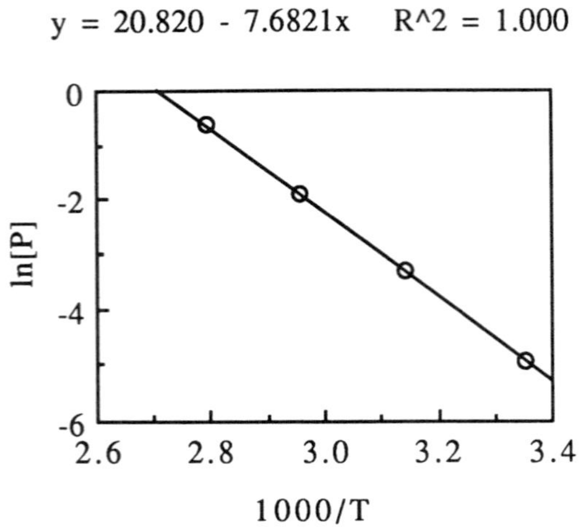

FIGURE 7.7 Data from Figure 7.6 plotted according to the Clausius-Clapeyron equation.

All three quantities will change with temperature,

$$\ln[P] = -\{\Delta H^{\P}/R\}(1/T) + \o'' \tag{7.6}$$

If K changes in a similar fashion, then this and Eq. 7.3 introduced into Eq. 7.5 give

$$\ln[S] = -\{\Delta H^{*}/R\}(1/T) + \o \; -\{\Delta H^{\P}/R\}(1/T) + \o'' =$$

$$\ln[S] = -\Delta H/RT + Q \tag{7.7}$$

where ΔH now is the heat of solution and Q is a combination of the ø-terms. It is pointed out that the heat of solution is not necessarily independent of temperature, and that K may not exhibit the temperature relationship shown above. ln[S] is, in such cases, not linear when plotted versus 1/T.

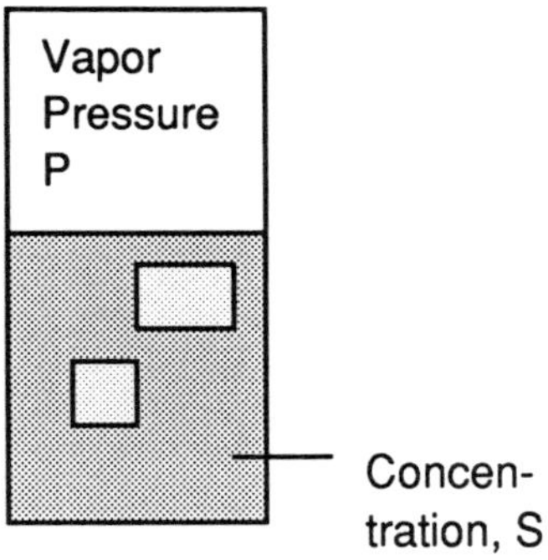

FIGURE 7.8 Vapor pressures in over-saturated systems.

7.5 SOLUBILITY MEASUREMENTS

Solubility has been treated as a concept in Chapter 2. Some aspects of it are, however, specific to polymorphism and are repeated here.

Determination of solubilities is carried out by placing an excess amount of solid in a liquid kept at constant temperature (e.g., in a constant temperature bath). The system is agitated for a sufficiently long period of time to establish equilibrium, i.e., until the drug concentration remains constant. There are, at times, problems due to hydrolysis of the drug in solution, but it is assumed here not to be the case. The solubility is a function of temperature and, in general, increases with temperature. (Exceptions to this are, for instance, methyl cellulose and many calcium salts.)

If solubility is measured versus temperature, then a curve such as shown in Figure 7.9 will result. An example of this (sulfanilamide in ethanol) is shown in the first two columns of Table 7.3.

TABLE 7.3 Solubility as a Function of Temperature

Temp (°C)	S (g/1000g)	1000/T	lnS
50	28	3.23	3.33
40	21	3.13	3.05
30	17	3.03	2.83
25	14	2.98	2.64

The relation that exists between temperature and solubility is:

$$\ln[S] = Q - [\Delta H/(RT)] \quad (7.8)$$

where S is solubility and ΔH is the heat of solution of the compound in the liquid. R is the gas constant. Data presented in this form from Table 7.3 are shown in Figure 7.10.

The two last columns in Table 7.3 show logarithms of S and 1000 times the inverse absolute temperatures. The least squares fit values are as shown: Slope = -2.572 (kCal/mole or if 1/T is used rather than 1000/T, then the unit is cal/mole), i.e.

$$\ln S = -2{,}572/T + 11.28 \quad (7.9)$$

The heat of solution is obtained from Eq. 7.8:

$$\Delta H/R = 2{,}572$$

so

$$\Delta H_{solution} = 1.99. \times 2{,}572 = 5{,}144 \text{ cal/mole} \quad (7.10)$$

It has been mentioned earlier that it is not always the case that $\Delta H_{solution}$ and this leads to non linear plots of ln[S] versus [1/T].

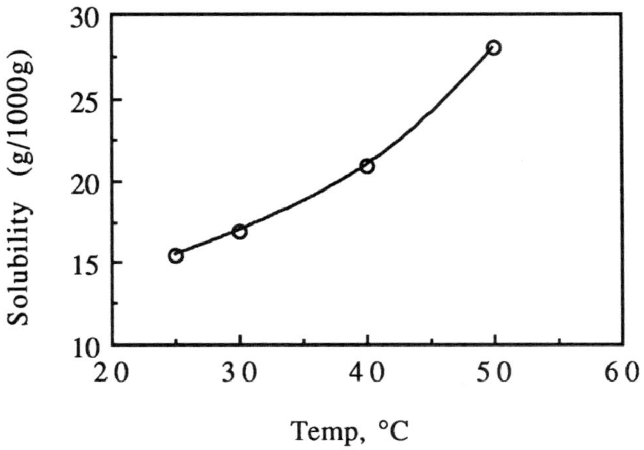

FIGURE 7.9 Solubility versus temperature curve.

Example 7.1

If only the first two points in Table 7.3 were used, what would the heat of solution be?

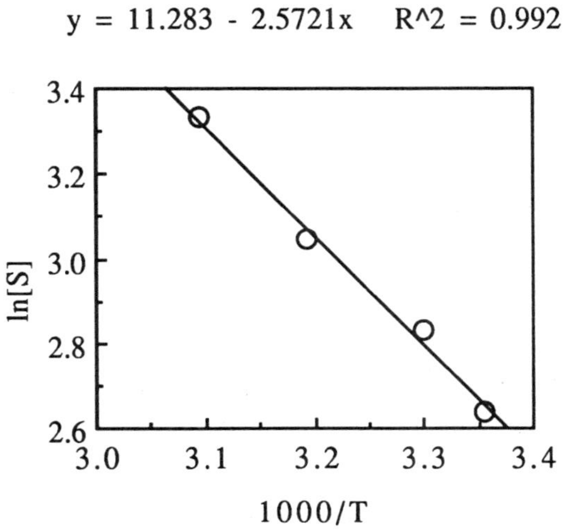

FIGURE 7.10 Semilog-inverse treatment of data in Table 7.3 (Eq. 7.8).

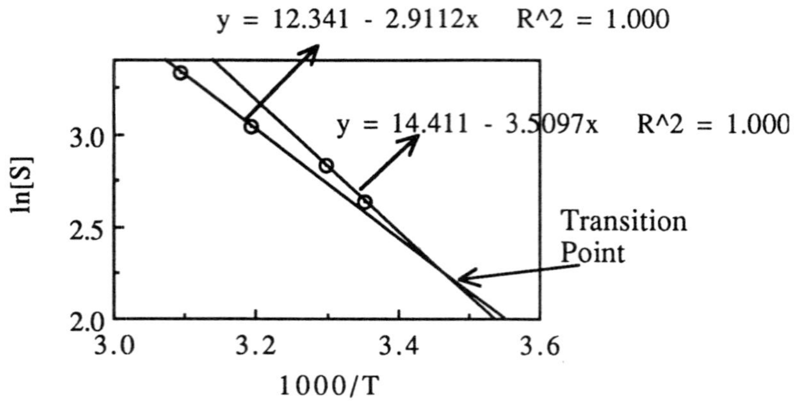

FIGURE 7.11 Example from Table 7.3.

Answer 7.1

The data are shown in a different fashion in Figure 7.11.

$\Delta H = 1.99. \times 2.911 = 5.8$ kcal/mole.

Example 7.2

If only the two last points in Table 7.3 are used, calculate the heat of solution.

Answer 7.2

From Figure 7.11 it follows that:

$\Delta H = 1.99. \times 3.509 = 7$ kcal/mole

7.6 POLYMORPHISM BY SOLUBILITY MEASUREMENTS

If the compound in Table 7.3 were recrystallized from ethanol, then an orthorhombic form would result, whereas if it were recrystallized from water, then a monoclinic form would result.

The two points at high temperature are actually the orthorhombic form and the two at lower temperature are the monoclinic form.

It is seen, therefore, that two lines result, and that these intersect in a so-called transition point (Point C in Figure 7.12). The intersection of the two lines is given by:

$$12.341 - (2{,}911.2/T) = 14.411 - 3{,}509.7/T$$

or

$$T = 598.5/2.07 = 289.13°K = 16°C$$

This means that above 16°C one form is more stable than the other. This type of polymorphism is denoted enantiotropy. Transition points obtained by solubility most often are of a slightly different nature as shall be discussed further below.

In this type of situation there is a form that is stable at room temperature and that has a vapor pressure of A (16 mm Hg, in the example). As temperature is raised slowly, at point C it will convert to another form (II) which on heating follows curve CE. The compound melts at point E. (The curve to the right of E is the vapor pressure curve of the melt.) If form I is heated rapidly from point A it may "pass through" C and show a melting point at D. On further heating it will either crystallize (since it is unstable, thermo-dynamically, with respect to II, which has a vapor pressure on the curve fraction CE, i.e., a lower vapor pressure), or it will simply stay liquid.

Another situation that may arise is that of monotropy, where there is no transition point, and where form II is metastable with respect to form I at all temperatures (Figure 7.13).

In this case the vapor pressure of form II is higher than that of form I at all temperatures and hence is not thermodynamically stable. It is denoted metastable, however, because as long as it is kept cool, dry, and of fairly large particle size, it will not transform.

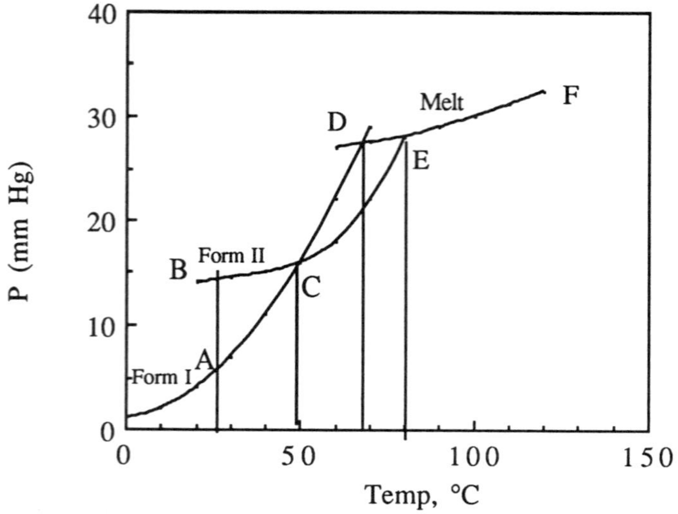

FIGURE 7.12 Enantiotropic polymorphic pair.

Most often, with enatiotropic pairs, the dissolution curve of any of the polymorphs as a function of temperature will be such that equilibrium has been reached at each temperature, and that, therefore, the solubility profile will reveal the transition point.

It is left to the reader to show that the transition temperature is at 40°C. Below it is assumed that the low temperature stable form is orthorhombic.

If the data from Table 7.4 are plotted as ln[solubility] versus inverse absolute temperature, a graph as shown in Figure 7.14 results. The "highest" line indicates the substance has a "higher" solubility than the form represented by the "lowest" line.

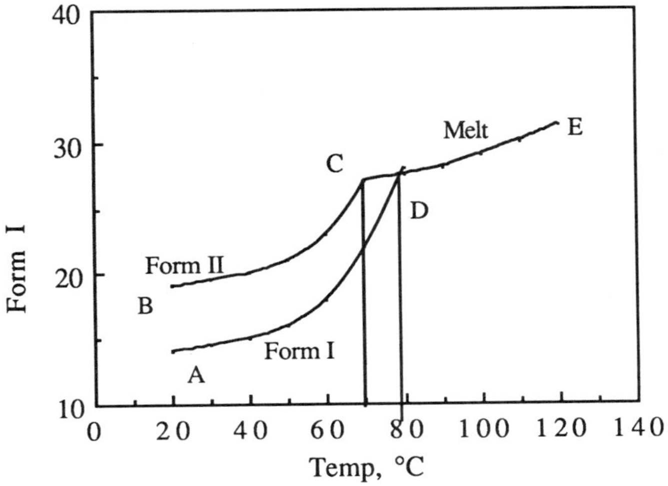

FIGURE 7.13 Monotropic pair.

Below 50°C the orthorhombic form is less soluble than the monoclinic form. The dissolved molecules do not differ. Above 50°C the monoclinic is less soluble. If the orthorhombic form is placed in water above 50°C, then the solution is supersaturated (so it is actually unstable) and will tend towards the stable system. Once monoclinic crystals start precipitating out, the

TABLE 7.4 Solubility of an Enantiotropic Solid

Temp,°C	1000/T	S (g/1000g)	ln[S]
20	3.411	1.105	0.1
30	3.000	1.350	0.3
40	3.193	1.649	0.5
50	3.095	1.822	0.6
60	3.002	2.014	0.7
70	2.914	2.226	0.8

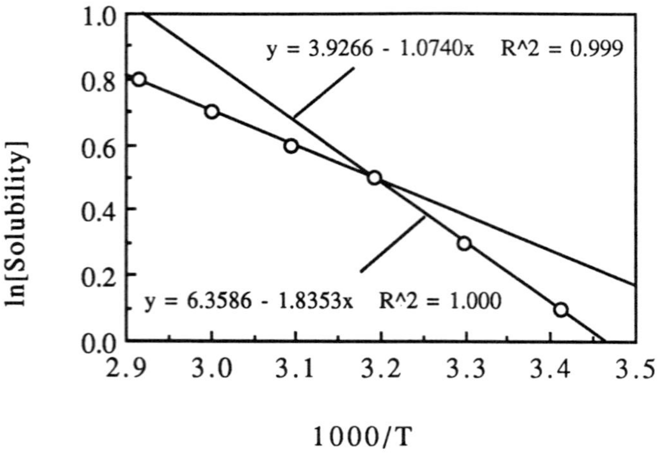

FIGURE 7.14 Data from Table 7.4.

system will convert to the stable (monoclinic) form (above 50°C). It should be pointed out that the supersaturated solution does not convert (apparently is stable) for some time. Therefore, if a polymorph has a higher solubility than another, it will dissolve more quickly since

$$\text{Dissolution rate} = k\ A\ (S\text{-}C) \tag{7.11}$$

so that if S is higher in one polymorph than in another, then the dissolution rate will also be higher.

7.7 METASTABLE SOLUTIONS

If water is let cool without agitation and very slowly it may be kept liquid below 0°C up until a certain point. Eventually ice will form. However, the metastable liquid state (or the glass as it is called) may exist for certain lengths of time.

If a solution of a substance when unsaturated is cooled down to where it just is saturated, then it will not precipitate. The temperature must be lowered further, so that precipitation occurs out of a supersaturated solution. The rate of precipitation is a function of the compound and of how "quiet" the system is kept.

If the metastable polymorph (above 50°C) is kept dry, it will not convert to the more stable form, and factors such as pressure (in tableting) or moisture (e.g. from wet granulation) are needed to cause a change of the crystal form.

When moisture is present then in time the polymorph will change to the stable form. This is exemplified in Figure 7.15. Moisture condenses on the solid (A). The moisture layer becomes saturated in the solid (S'), but this solubility is higher than that of the stable form (S). In time, precipitation of the stable form (B) will occur. The crystals of the stable form will experience a solution that is supersaturated and hence will further precipitation. At this stage (C) the metastable solid remaining will experience a milieu which is unsaturated and hence will dissolve (C) and this will continue until all is dissolved (D), i.e., the conversion is complete.

Eq. 7.8 is of advantage pharmaceutically, if the transition point is close to 37°C, i.e. body temperature. The more soluble form at 37° will give the fastest in vivo dissolution, and it is therefore of importance to establish solubilities of various crystal forms at different temperatures. The principle of this has been demonstrated above.

In pharmaceutical history there have been cases where this has been important. Chloramphenicol is a case where the innovators' product appeared to be more effective than the generic products. It was shown later that this was due to a metastable polymorph with higher solubility with higher dissolution rates. The FDA, upon this, established a policy of dissolution testing in all ANDA's.

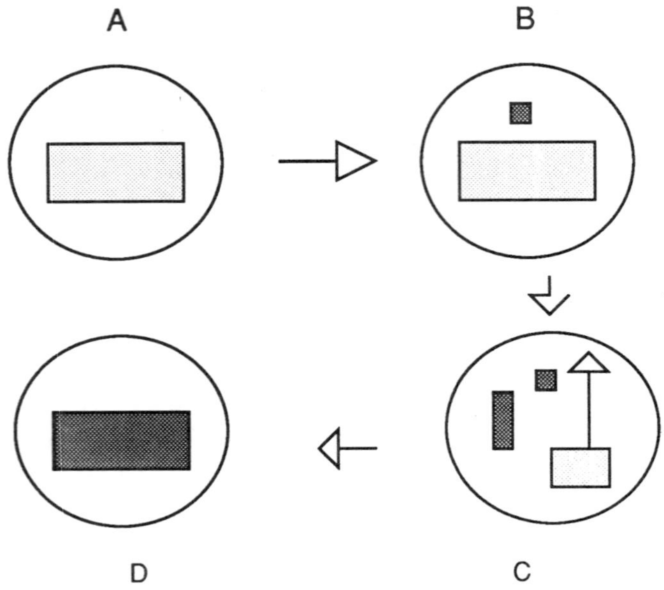

FIGURE 7.15 One mechanism whereby a metastable polymorph reverts to the stable species.

There are obviously cases where polymorphism is NOT of significance, such as cases where the transition point is about 37°C because then the solubility difference is small. An example is ampicillin trihydrate. Therefore, the existence of polymorphism in itself is not a sufficient reason to rule out use of a different polymorph in generic equivalents.

The form that is metastable at 25°C (room temperature in the U.S.) should NOT be employed in the case of ready-made suspension formulations. The form will invariably revert to the stable form on standing, thus giving rise to crystal growth. It should be pointed out that the amorphous form of a drug substance, in most every case, will be more soluble than its crystalline form (whether the crystalline form be metastable or not).

7.8 DIFFERENTIAL SCANNING CALORIMETRY (DSC)

Companies prefer to market dosage forms using the polymorph which is stable at 25°C, so preformulation efforts are designed to concentrate on this.

X-ray studies are usually used but DSC may also be employed. A simple DSC trace (demonstrating a melting point endotherm) is shown in Figure 7.16. Endotherms, as shown here, register up. (On some DSC instruments it registers in downward direction, but the up-convention is employed here.)

If the compound exhibits enantiotropic polymorphism, then the trace may be such as shown in Figure 7.17.

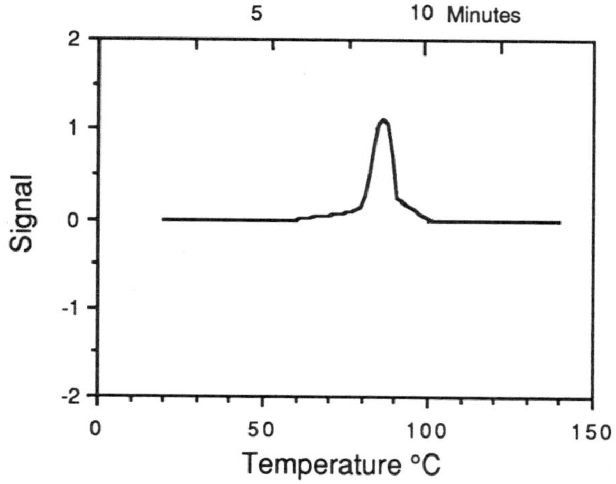

FIGURE 7.16 DSC trace of a compound melting at 86°C. Heating rate is 10°C.

Monotropic polymorphic forms exhibit the type of trace shown in Figure 7.18. There are times when the recrystallization exotherm is missing from the metastable profile.

In this case (if the compounds are stable to melting) recooling and reheating will develop the melting DSC trace for the stable modifications (Figure 7.16).

Recent literature is replete with competent work dealing with the subjects under discussion. Good examples are the work of Joszwiakowski et al (1996) regarding lamivudine and the work of Ledwidge et al., (1996) on the subject of diclofenac. Chan and Gonda (1995) have reported on the physicochemical

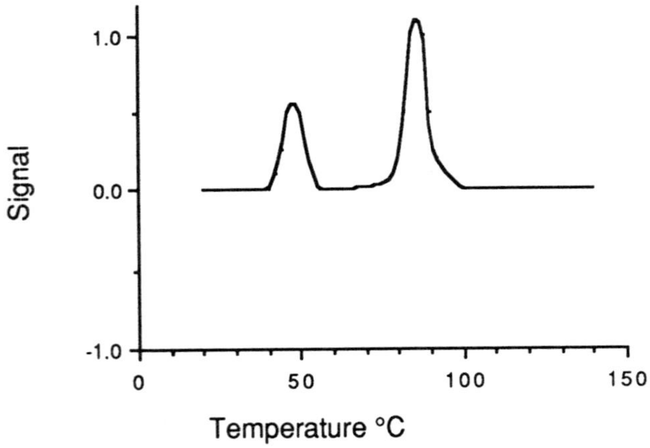

FIGURE 7.17 DSC of an enantiotropic compound. The transition temperature is 50°C and the melting point the same as in Figure 7.15.

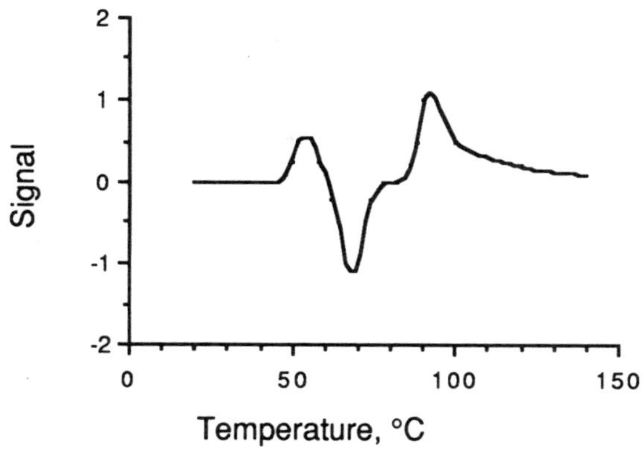

FIGURE 7.18 DSC profile of a monotropic polymorph, with a melting temperature of 50°C. Recrystallization (to a stable polymorph) takes place immediately after melting, whereupon the stable modification melts at about 90°C.

characterization of nedocromil, Otsuka et al. (1995) have reported on chlorpropamide polymorphs, and Ghosh et al. (1995) have reported on crystal structures and thermal properties of dialkylhydroxypyridones.

7.9 PREFORMULATION TASKS

Since polymorphism is so important, the following should be routine in preformulation of solid, neat drug substances: (a) DSC traces and (b) X-ray powder diffraction should be carried out. (c) Solubility determinations should be carried out. (d) These are carried out in the spirit of dissolution testing, since, if a metastable compound is at hand, a larger solubility will first be reached, which will then decrease to asymptote at the "stable" solubility.

REFERENCES

Chan, H.-K.. and Gonda, I., (1995), *J. Pharm. Sci.,* **84:** 692.

Ghosh, S., Ojala, W.H., Gleason, W.B. and Grant, D.J.W., (1995), *J. Pharm. Sci.,* **84:**1392.

Jozwiakowski, M. J., Nguyen, N.-T., Sisco, J. J., and Spankcak, C. W., (1996), *J. Pharm. Sci.,* **85:**193.

Ledwidge, M. T., Draper, S. M., Wilcock, D. J., and Corrigan, O. I., (1996), *J. Pharm. Sci.,* **85:**16.

Otsuka, M., Matsumoto, T., Higuchi, S., Otsuka, K., and Kananiwa, N., (1995), *J. Pharm. Sci.,* **84:**614.

CHAPTER 8

Drug Substance Considerations

Preformulation often deals with problems arising from the synthesis (purification) of the drug substance. The synthesis also may have an effect on polymorphism, an important topic reserved for a subsequent chapter.

So both for the innovator and the generic manufacturer it is important to control the physical state of the drug substance. The method of purification of the drug may change the drug substance's physical characteristics to such a degree that its machinability (i.e. the capability of making it into tablets or capsules) would be impaired.

8.1 SALT SELECTION

Many drugs are marketed as specific salts (amine hydrochlorides, sodium salts) and the user might wonder why certain salts are used rather than the corresponding amine or acid. A few of these reasons are:

(a) The acid may not be a solid but rather be an oil.
(b) Salts of acids are most often more soluble than the corresponding acids.
(c) They usually crystallize out more easily in the recrystallization (purification) step.

Clavulanic acid (an inhibitor of ß-lactamase used in Augmentin, SKB) is an oil. The potassium salt, however, is well defined. (It happens to be highly soluble; this point will be discussed further at a later point.)

Excessive solubility may at times be a drawback. Although higher solubility would in general imply equal or better bioavailability than that of a less soluble modification, excessive solubility gives rise to more pronounced hygroscopicity. This, also, will be described at a later point in the text.

8.2 PURIFICATION BY PRECIPITATION

It was mentioned in the previous chapter that purification is the last step of a synthesis of a drug substance. Thermal recrystallization is often used, but precipitation is also a common method of achieving purity.

Precipitation may be accomplished by either changing the pH of a protolytic substance or by use of mixed solvents. *Precipitation* is a way of purification which is best described by an example. The solubility of a drug substance in acetone is 132 g per 1000 g of solvent at 10°C. If water is added then the solubility will decrease as shown in Figure 8.1.

If 120 g of drug substance are dissolved in 1000 g of acetone (Figure 8.1) and if 2000 g of water are added (to render the concentration of acetone 0.33), then the solubility decreases to 10 g per 1000 g (as shown in Figure 8.2), i.e. 30 g per 3000 g of solvent. Hence 132 minus 30, in other words 102 g, of pure compound will precipitate out. The purity will be a function of the level of impurities in the compound before reprecipitation.

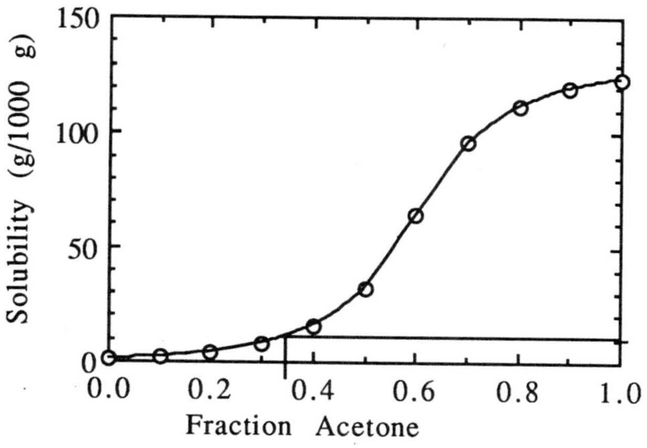

FIGURE 8.1 Solubility of a compound in mixtures of acetone and water.

A good example of solubility in mixed solvents and its potential for precipitation purification is the work by Joszwiakowski et al. (1996). Precipitation purification requires strict testing for residual solvent.

8.3 PURITY ASSESSMENT BY MEANS OF PHASE SOLUBILITY

Purity may be assessed by determining a substance's melting point, T. If a compound contains a mole fraction, x, of an impurity, and the melting point of the "pure" melting temperature is denoted T_o°K, then the "impure" melting point, T°K is given by:

$$\ln[1-x] \approx -x = -\Delta H/R\{(1/T)-(1/T_o)\} \qquad (8.1)$$

This may be rewritten

$$x = \Delta H/R\{\Delta T/T_o^2\} \quad (8.2)$$

where R is the gas constant and ΔH is the heat of fusion.

The melting points may be obtained by thermoanalytical means (DSC). In so doing the entire "eutectic curve" may be constructed.

Example 8.1

If ΔH equals 8,000 cal/mole and To were 200°C, construct the melting point curve by means of Eq. 8.1.

Answer 8.1

T may be calculated for a series of x-values. A program to carry out this is shown below. Eq. 8.1 may be written:

$$(1/T) = (1/T_o) - (R/\Delta H)\ln(1-x) \quad (8.3)$$

ΔH = 8000 and T_o = 200 when entered give the results in Table 8.2.

TABLE 8.1 Program for Eq. 8.3

```
100 INPUT "Heat of Fusion, (Positive, cal/mole)=";H1
110 INPUT "Melting Point, °C="; T1
120 PRINT "Mole Fraction", "Melting Point, °C"
130 FOR X1 = 0 TO 1 STEP .1
140 T2 = T1 + 273.15
150 T3 = 1/T2
160 X2 = 1.99/H1
170 X3 = LOG(1-X1)
180 X4 = X2*X3
190 X5 = T3-X4
200 X6 = 1/X5
210 X7 = X6-273.15
200 PRINT X1,X7
210 NEXT X1
```

TABLE 8.2 Results from Table 8.1 using ΔH = 8,000 cal/mole and T_o = 200 °C

Mole Fraction	Melting Point, °C
0	200
0.1	196.4
0.2	192.4
0.3	188.0
0.4	183.1
0.5	177.4
0.6	170.8
0.7	162.5
0.8	151.5
0.9	134.3

If the value of $\Delta H/R$ is back-calculated from Figure 8.2, the value is 4.0216 or, by multiplying by R = 1.99, 8.0 kCal/mole as it should be.

The curve for this is shown in Figure 8.2.

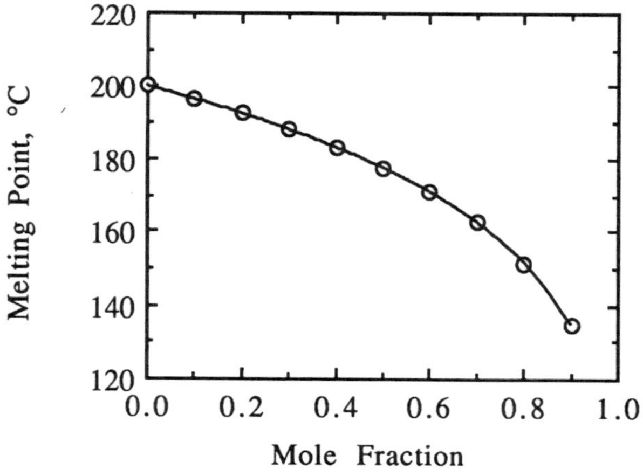

FIGURE 8.2 Data from Table 8.2.

8.4 DIFFERENTIAL SCANNING CALORIMETRY (DSC)

The most common way to assess impurities in preformulation is by the use of DSC. In this method a sample of the drug substance is transferred to a sample holder. This is heated at a given heating rate. The heat flow is measured and compared to that of a reference (with a given and known heat capacity). If a heat event occurs (e.g. melting), an endothermic peak will result. The magnitude of this (e.g. heat of melting) is directly proportional to the area between the curve and the baseline. This is quantitated by comparison with the heat of fusion of indium, which is employed as a standard of reference.

If a set of data are obtained as shown in Table 8.3, it is possible to use this as a calibration curve for future reference and to calculate the heat of fusion.

The data have been plotted in Figure 8.3, and it is seen that the heat of fusion is 1.99 x 4.25 = about 8.5 kcal/mole.

The data in Table 8.3 if extended beyond 85°C would have been extended further than reality. If the impurity is denoted B, and supposing that pure B had a melting point of 140°C, then adding a little of A to B would result in a melting point depression of B. Hence the complete phase diagram would be as shown in Figure 8.4.

TABLE 8.3 Example of Data Treated by Eq. 8.1

Temp,°C	Temp, T, °K	1000/T	(1-x)	ln(1-x)
102.000	375.150	2.666	1.000	0.000
98.400	371.550	2.691	0.900	-0.105
94.400	367.550	2.721	0.800	-0.223
90.000	363.150	2.754	0.700	-0.357
85.100	359.250	2.784	0.600	-0.511

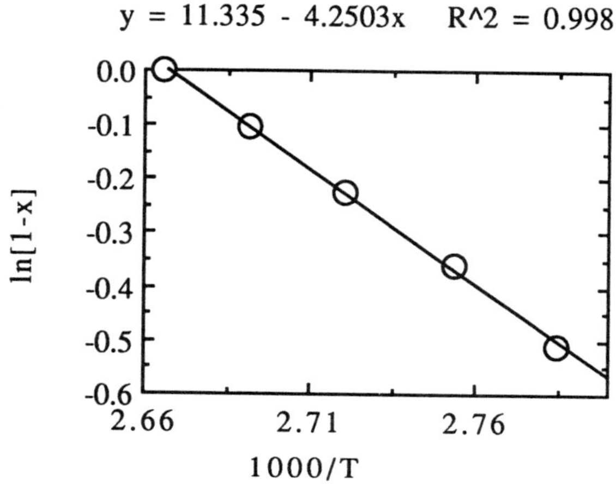

FIGURE 8.3 Data from Table 8.4 treated by Eq. 8.1.

8.5 EUTECTIC DIAGRAMS

Referring to Figure 8.4, at temperatures below C (86°C) the system is solid, regardless of the amount of B in A. In the triangular area ACE there are two phases, solid A and a liquid of the composition given by the line AE. For example, if a solid containing 0.2 fraction of B were heated, it would give signs of melting at 86°C and then completely melt at 95°C.

Conversely the triangular area EBD corresponds to a two phase system of compound B and a liquid the composition of which is given by the liquidous line EB.

Above the line AEB there is only one phase (i.e., liquid).

Eutectic diagrams may be established by means of DSC. Figure 8.5 shows a schematic DSC trace, in which the eutectic temperature is 98°C and the temperature on the liquidous line (the "melting point") is 102.5°C.

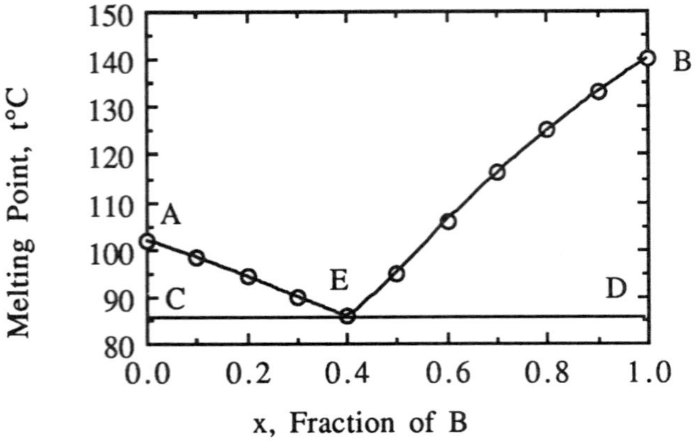

FIGURE 8.4 Eutectic diagram of two compounds, A and B.

If samples are made of different ratios of A and B, then the onset of the signal will be the same at all mixtures (provided it is a simple eutectic), and the point where the signal returns to base will be the point on the liquidous line. A (simple) eutectic diagram can be established in this fashion.

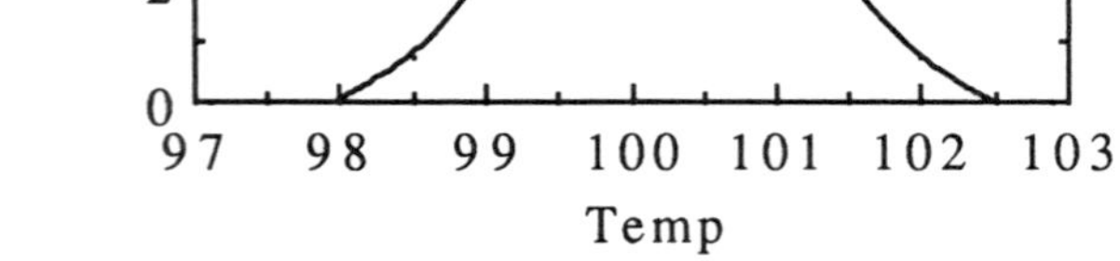

FIGURE 8.5 Example of DSC trace of impure solid. The eutectic temperature is the lowest point of the trace.

It should be noted that when water is the "left" component, the diagrams is of the type used for lyophilization studies, and in this case it is of importance to establish the vacuum drying cycles.

If water is the "left" component, then the right half of the diagram constitutes the solubility/temperature curve of the "right" compound in water (Carstensen, 1977, Denbigh, 1961).

8.6 PREFORMULATION TASKS

The preformulator should obtain, with reference to the drug substance raw material

(a) X-ray diffraction pattern

(b) DSC scanning up and down

(c) TGA (to test for bound and unbound moisture)

(This substitutes the usual loss on drying)

(d) pH of a slurry or a 1% solution of the drug substance

(e) Eutectic diagrams with mixtures with different, pertinent excipients that may apply to the system under investigation (e.g., a tablet).

REFERENCES

Carstensen, J.T., (1977), *Pharmaceutics of solids and solid dosage forms,* John Wiley and Sons, New York.

Denbigh, K., (1961) *The principles of chemical equilibrium,* Cambridge University Press, p. 264

Jozwiakowski, M. J., Nguyen, N.-T., Sisco, J. J., and Spankcak, C. W., (1996), *J. Pharm. Sci.,* **85:**193.

CHAPTER 9

Moisture Isotherms

Before discussing the actual tie-in of preformulation with formulation of solid dosage form, it is necessary to describe a few qualities of solids that are of importance in solid dosage form development.

9.1 WATER ADSORPTION

Water adsorption deals with the extent and rate with which water is adsorbed onto surfaces. If a sample of a substance is placed in a desiccator with a given relative humidity (RH), then, given the circumstances, it will pick up moisture.

In studying water adsorption it is, therefore, necessary to have a means of creating a space with a given relative humidity. One way of achieving this, as shall be seen in the following, is to place a saturated solution in the bottom well of a desiccator and to place the drug substance (or excipient) on a Petri dish on the screen of the jar. A saturated solution of, for instance, an electrolyte has a given relative humidity at a given temperature. To make sure that the solution indeed is saturated, it is customary to have an excess of the electrolyte present, so that as long as one can observe crystals in the bottom of the desiccator one may be sure that the RH is the desired value.

The air mass in the desiccator is fairly stagnant and this is a disadvantage; more modern approaches overcome this deficit.

The manner in which the experiment is carried out is to place the sample of drug substance (or excipient) in a tared, open Petri dish, laying the cover next to it. Fom time to time the desiccator is opened, the cover is placed on the Petri dish, it is weighed, returned, the cover removed, and the desiccator closed again. The manipulation is necessary because the drug substance may pick up moisture in transfer from desiccator to balance. There are usually zero time problems, but shortly after onset the weight gain will be linear in time. The rate is divided by the initial rate, and this rate (slope) is referred to as the *moisture uptake rate (MUR).* To make the experiment more

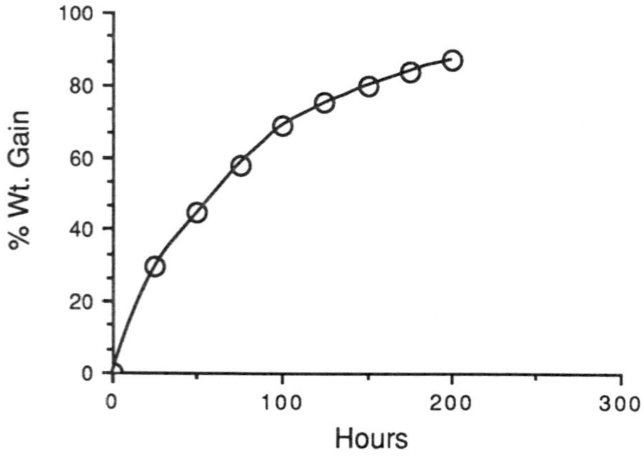

FIGURE 9.1 Data for choline chloride moisture adsorption at 100% RH. Drawn from data published by VanCampen et al. (1980).

accurate, it should be carried out in a dry box of the same relative humidity as the experiment, but this is usually not done. Figure 9.1 is an example of MUR curves.

9.2 CLASSIFICATION OF SOLIDS IN REGARDS TO MOISTURE

From the preformulator's point of view there are three types of moisture isotherms:

(a) solids that form no stoichiometric hydrates and are not water-soluble. Examples of this are starch and microcrystalline cellulose.

(b) solids that form no stoichiometric hydrates but are soluble to some measurable extent. An example of this would be sucrose.

(c) solids that form stoichiometric hydrates and are soluble to some measurable extent.

9.3 CRITICAL MOISTURE CONTENT (RH*)

Type (b) solids will be discussed first. A non-hydrating solid will pick up moisture when placed in an atmosphere of higher than zero percent RH. At low RH the adsorption will be a surface adsorption, but once a given RH value (the critical RH) is reached, then moisture will adsorb more rapidly. The critical RH-value is the water vapor pressure, P_s, above a saturated solution of the compound (Yoshioka and Uchiyama, 1986a, 1986b).

If a solid is placed in an atmosphere that has a vapor pressure, P_a, that is higher than the vapor pressure, P_s, of the saturated solution of the compound, then moisture will condense upon the solid. This is demonstrated in Figure 9.2.

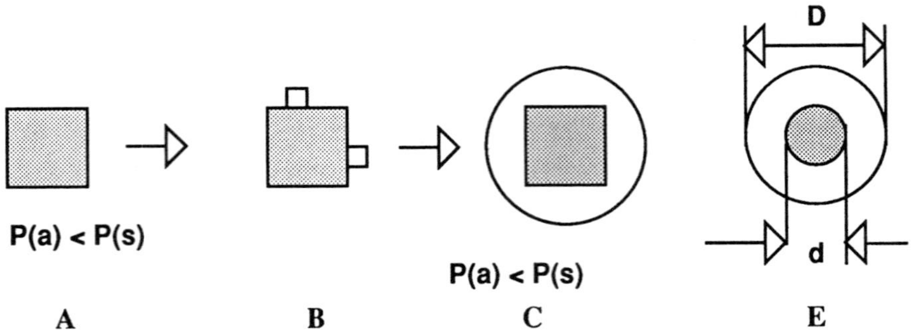

FIGURE 9.2 Schematic of water adsorption (A) and (B) below and (C) above the critical RH. (E) is the same representation as (C) except in spherical presentation.

The condensed water will dissolve solid drug, and it will be assumed below that the sorbed solution is saturated at all times.

The question is, what sort of curve might be expected for the extent of moisture uptake with time (the moisture uptake rate curve, the MUR curve)?

Assume that, at time, t, a certain amount of moisture, w, has been adsorbed by a particular solid particle weighing m grams and of dimension (e.g., diameter), d_o, where the subscript denotes the condition prior to moisture adsorption. At time, t, moisture will have adsorbed, some solid will have dissolved, and the diameter, d_s, of the solid itself will have decreased from its original value, d_o. As is obvious from Figure 9.2(e) , the diameter of a spherical particle, D, is the sum of the diameter, d, of the remaining, undissolved solid drug, and the thickness, h, of the moisture layer, i.e., D = d+h. Since d changes with time, it is denoted d_s in the following.

The profile for one drug particle will first be considered. It is assumed that the particle is spherical and has an original diameter of d_o, and an original mass of m_o and the solids density is ρ_s. Hence:

$$m_o = \rho_s \pi d_o^3/6 \qquad (9.1)$$

At time, t, there will be a certain mass, w, of water adsorbed. The drug substance has a solubility of s g/g in the sorbed water. The amount of drug present in solid form at time, t, is then given by the original amount less the amount dissolved, i.e.:

$$m = m_o - ws \qquad (9.2)$$

The diameter, d_s, of the solid drug left at time t, is therefore given by:

$$\rho_s(\pi/6)d_s^3 = m_o - ws \qquad (9.3)$$

or $$(\pi/6)d_s^3 = [m_o - ws]/\rho_s \qquad (9.4)$$

As mentioned, the liquid layer is assumed to be a saturated solution of the drug in water at all times. The volume fraction of drug in the saturated solution is denoted x_s, and the densities of solid and water, respectively, denoted ρ_s and ρ_w. Assuming ideality, the density of the saturated solution, ρ^*, will be given by:

$$\rho^* = (1-x_s)\rho_o + x_s\rho_\sigma \qquad (9.5)$$

Hence the volume, v_l, of the sorbed liquid is given by:

$$v_L = (w+ws)/\rho^* = w(1+s)/\rho^* \qquad (9.6)$$

Since the layer is assumed saturated at all times, its density will be time-independent, and equal to ρ^*.

The diameter of the particle (i.e., the solid drug plus the liquid layer) at time, t, is denoted D in the following, so that the volume of the entire particle, v_T, is:

$$v_T = v_L + v_S \tag{9.7}$$

or $$(\pi/6)D^3 = w(1+s)/\rho^* + (\rho\pi/6)d^3 \tag{9.8}$$

Introducing Eq. 9.4 into this gives:

$$(\pi/6)D^3 = w(1+s)/\rho^* + [m_o - ws]/\rho_S \tag{9.9}$$

or

$$D^3 = (6/\pi)[w(1+s)/\rho^* + \{(m_o - ws)/\rho_S\}] \tag{9.10}$$

This may be written:

$$D^3 = (6/\pi)[w\{((1+s)/\rho^*)-(s/\rho_S)\}+ (m_o/\rho_S)] = \tag{9.11}$$

In general the coefficient to w (the term "$\{((1+s)/\rho^*)-(s/\rho_S)\}$") will be negative because

$$\{(1+s)/\rho^*) < (s/\rho_S\} \tag{9.12}$$

when

$$1+s < (\rho_S/\rho^*)S \tag{9.13}$$

or when

$$s[1-(\rho_S/\rho^*)] < 1 \tag{9.14}$$

Since the term $[1-(\rho_S/\rho^*)]$ is less than one (except in rare cases), s would have to be more than one, and the more it is the larger the value of (ρ_S/ρ^*). The derivation to follow can be carried out regardless of the sign of the coefficient to w, but is, in the following, assumed to be negative. Eq. 9.11 can, therefore, be written:

$$D^3 = (6/\pi)A[B-w] \tag{9.15}$$

where

$$A = [(s/\rho_s) - ((1+s)/\rho^*)] \tag{9.16}$$

and

$$B = (m_o/\rho_s)/[(s/\rho_s) - ((1+s)/\rho^*)] \tag{9.17}$$

The condensation equation states that the condensation rate is proportional to the pressure differential, P_a - P_s, where P_a is the water vapor pressure in the atmosphere and P_s is the water vapor pressure above a saturated solution of the drug in water. It is also proportional to the external area of the particle (πD^2), and the proportionality constant, k, is a mass transfer coefficient. Hence:

$$dw/dt = k(\pi D^2)(P_a - P_s) \tag{9.18}$$

Introducing Eq. 9.15 yields

$$dw/dt = k\pi\{(6/\pi)A[B-w]\}^{2/3}(P_a - P_s) \tag{9.19}$$

or

$$dw/dt = 3G[B-w]^{2/3} \tag{9.20}$$

where

$$3G = k\pi\{(6/\pi)[A^{2/3}]\}(P_a - P_s) \tag{9.21}$$

At given atmospheric conditions (constant RH) G is a constant. Eq. 9.21 integrates to:

$$[B-w]^{1/3} = [B]^{1/3} - Gt \qquad (9.22)$$

where the initial condition that w = 0 when t = 0 has been imposed. Eq. 9.22 can be solved by iteration.

If G is calculated from such plots carried out at different atmospheric conditions, then it can be plotted versus RH to give a straight line and P_s can be calculated from the intercept to slope ratio, and kA can be calculated from the slope.

If a sample of solid drug is monodisperse, with N particles per gram, then:

$$N(\pi/6)d_o{}^3\rho_s = 1 \qquad (9.23)$$

Multiplying through Eq. 9.22 by $N^{1/3}$ gives the same equation, where w is now weight of moisture adsorbed per gram of dry solid, and G and B hence are increased by the same factor. Multiplying through by $100^{1/3}$ then gives the same in percent moisture uptake (calculated on the basis of dry solid), so that the format of Eq. 9.22 is the same, regardless of the unit of uptake used. The values of G and B must, of course, be adjusted accordingly (Carstensen et al., 1987).

An example is shown in Table 9.1.

TABLE 9.1 Moisture Uptake of a Compound at 80% RH

Time (days)	w	$(35-w)^{1/3}$	$(29-w)^{1/3}$
0	0	3.233	3.038
1	15	2.687	2.389
2	22	2.331	1.901
3	25	2.138	1.580
4	27.5	1.944	1.143
5	28	1.855	0.796

Using various values of B as an iterant gives the curves shown in Figure 9.3, and it is seen that the value of B = 29 linearizes the data. The correlation coefficient, as shown, is good.

9.4 MOISTURE UPTAKE BELOW THE CRITICAL RH

Type (a) solids and type (b) solids below the critical relative humidity will be discussed next. As shown in Figure 9.2, at RH values below the critical RH, only small amounts of moisture will be adsosrbed. This is surface adsorption, and if moisture (weight) gain studies are carried out at relative humidities below the critical RH, then weight gain curves as shown in Figure 9.4 will result.

If the "equilitrium" levels are plotted as a function of RH, then curves as shown in Figure 9.5 result. These curves are denoted Brunauer-Emmett-Teller or BET curves (Brunauer et al., 1938, 1959).

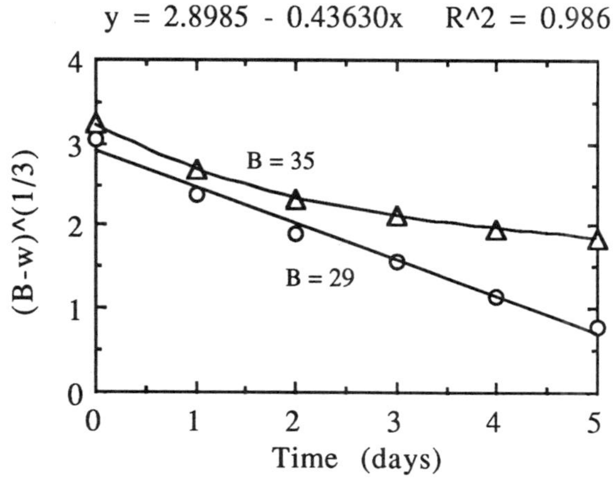

FIGURE 9.3 Data from Table 9.1 treated by Eq. 9.22

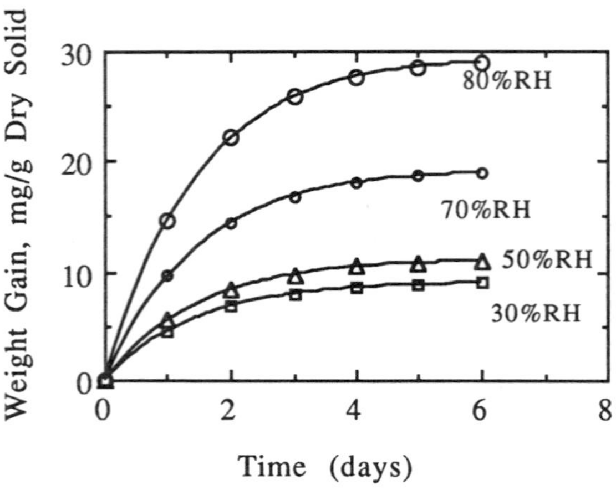

FIGURE 9.4 Water adsorption of a compound as a function of RH. The RH-values studied are below the RH over a saturated solution of the compound. The curve at 80% RH is the same as treated in Table 9.1 and Figure 9.3.

If, at a given temperature, T°K, and water vapor pressure, P, the volume (at STP) adsorbed is V per gram of dry solid, then the BET equation states that:

$$P/\{V(P_o-P)\} = (1/\{V_m c\}) + (\{c-1\}/\{V_m c\})(P/P_o) \tag{9.24}$$

where P_o is pure water's vapor pressure at T, and where V_m is the volume of water adsorbed in a monolayer, i.e. just covering one gram of solid. (Limitations on this statement will be covered below.) In the above equation

$$c = (H_1 - H_L)/RT \tag{9.25}$$

where H_1 is the heat of adsorption in the first layer and H_L is the heat of condensation of water. When c is large, then Eq. 9.24

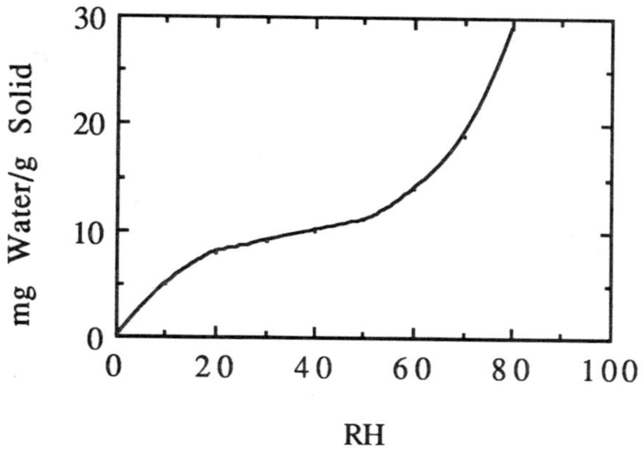

FIGURE 9.5 BET isotherm from the data in Fig. 9.4. The compound forms no stoichiometric hydrates and is exposed only to relative humidities below that of a saturated solution of the compound.

becomes:

$$V = \beta/(100\text{-RH}) \qquad (9.26)$$

where ß is a constant.

There are, for the purpose of preformulation, three types depending on the type of compound that is being studied. If it is not "water-soluble" then the moisture will adsorb in layers. In fact the BET equation is derived on this assumption, assuming that H_1 is associated with the first layer and H_L is associated with the following layers. So there will be a first layer, but before that is filled up, a new layer will start, and this will continue.

If, after 90% RH (or some other high relative humidity) is reached in an adsorption experiment, the process is reversed, i.e., the RH is reduced to, say, 80% RH, then one might expect that the amount of moisture remaining would be the same as on the way "up." This is not the case, and in most instances a different desorption curve results (Figure 9.6). This curve is *metastable*; i.e., it is reproducible, but it is not thermodynamically stable. In this writing, except where otherwise noted, it will be considered stable.

9.5 AUTOMATED EQUIPMENT

The moisture uptake experiments may be carried out in an automated vacuum balance, where the weight *of a small sample* is monitored until the weight increases by less than a preset amount (e.g. 0.01 mg) in a certain time period. The relative humidity then automatically increases by a certain increment,

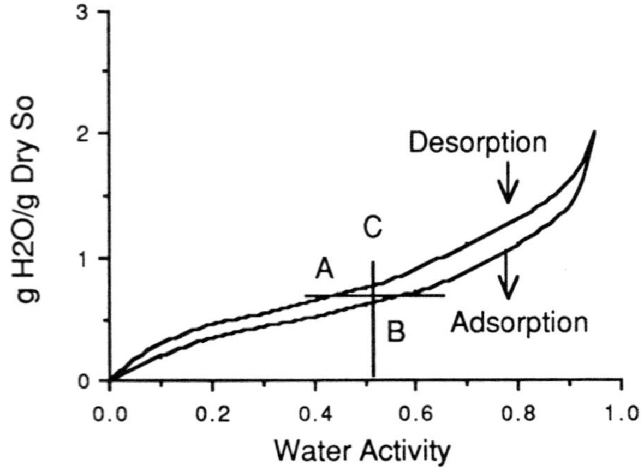

FIGURE 9.6. Sorption and desorption (up and down) isotherm of a solid that forms no hydrates and is "water-insoluble."

and this is continued until 90%RH is reached. The process is then reversed to get the *desorption* isotherm. An example of this is shown in Figure 9.6.

The problem with this approach is that *attainment of true equilibrium is exceedingly slow at times*, and many reported isotherms are in error for this reason. In the case of some hydrates, for instance, it has taken more than a year for equilibrium to be established, and the error committed in these cases, had only a couple of months, let alone an hour been used, would have been considerable.

The advantage with automated vacuum balances is that they are very sensitive, and very small samples can be used, and *small samples equilibrate more rapidly.*

9.6 ISOTHERMS OF "SOLUBLE" COMPOUNDS

If the compound is soluble to some extent, then the isotherm will be as in Figure 9.5 up to the point where the vapor pressure equals that of the saturated solution.

This will result in an isotherm as shown in Figure 9.7 where the vapor pressure over a saturated solution is assumed to be 50% RH. At this RH the compound will "adsorb" sufficient water to form a saturated solution (section AB).

The part of the isotherm to the right of A is based on the ideal solution equation that:

$$P/P_o = a = (1-x) \tag{9.27}$$

where x is mole fraction and is given by

$$x = (V/18)/\{(1/Q)+(V/18)\} = V/\{(18/Q) + V\} \tag{9.28}$$

Q is here the molecular weight of the compound. Inserting Eq. 9.27 into Eq. 9.28 gives:

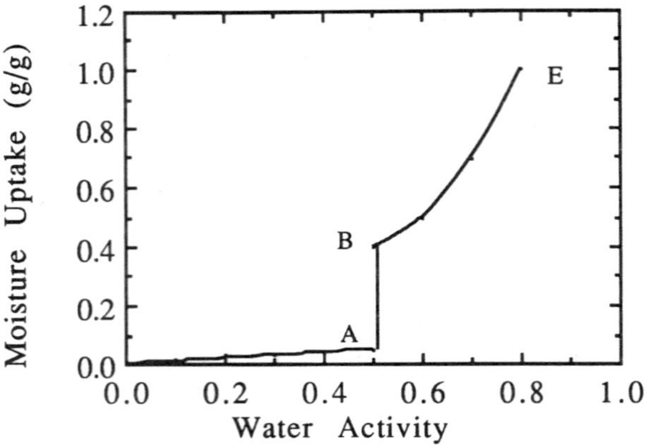

FIGURE 9.7 Isotherm of a solid that forms no stoichiometric hydrates but is soluble to some measurable extent.

$$a = 1 - V/\{(18/Q) + V\} = q/\{q + V\} \tag{9.29}$$

where $q = 18/Q$. Eq. 9.29 rearranges to:

$$V = \{q/(1+q)\}/a \tag{9.30}$$

The BE part of the isotherm follows Eq. 9.30.

9.7 HYDRATES

Many compounds form hydrates, i.e., for a monohydrate:

$$A + H_2O <-> A \,.\, H_2O \tag{9.31}$$

An example of an isotherm for such a compound is shown in Figure 9.8. There are several characteristics about it. Below a given RH (in this case 30%) the compound only picks up moisture to the extent of surface adsorption. If exposed to an

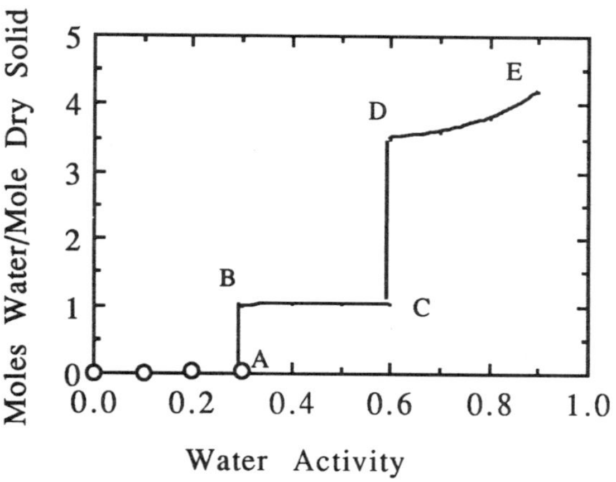

FIGURE 9.8 Isotherm of a solid that forms stoichiometric hydrates and is soluble to some measurable extent.

atmosphere of 30% the compound will pick up one mole of water. If exposed to atmospheres above 30% (but less than 60%RH), the hydrate will pick up only surface moisture. If the hydrate is exposed to an atmosphere of 60%, then it will pick up an amount of water that will transform it into a saturated solution. (In the example this is an additional 2.5 moles of water, i.e. the solubility of the compound would be 1 mole/3.5 moles of water.) Beyond 60% the compound will pick up sufficient water to create an unsaturated solution.

It is noted that a mixture of anhydrate and monohydrate will create an atmosphere of 30%, i.e., any mixture of the two will be stable at exactly this relative humidity. If the monohydrate is placed in a bottle that is completely dry, and the bottle then capped, then a small amount of water of hydration will evaporate to create 30% in the bottle, and the compound will therefore be (ever so slightly) dehydrated, but it will principally remain in the form of the monohydrate. (This, of course, depends on the volume of the headspace).

Between 30 and 60% RH, in this example, the hydrate will be stable (in the sense that it does not dehydrate), and often hydrates are stored in desiccators over a salt solution with a relative humidity in between that of the hydrate and its saturated solution.

9.8 HYGROSCOPICITY

Hygroscopicity of a solid is the rate and extent to which it adsorbs water. This is a function of (a) the surface area of the powder, (b) the differential between the vapor pressure in the atmosphere and the equilibrium vapor pressure of the compound, and (c) a kinetic term, k.

$$dM/dt = kA(P_{atm}-P_{eq}) \qquad (9.32)$$

The initial slopes of the curves in Figure 9.4 (or the part where the curve is linear) may be plotted as a function of relative humidity. These moisture uptake curves will give uptake rates that are linear in relative humidity, and the point where they become zero will be the equilibrium RH of either the hydrate salt-pair (point C in Fig. 9.9) or the saturated solution (the intersection with the x-axis in Fig. 9.10).

In desiccator work, this is a practical way of obtaining solubility, because if the moisture uptake rate is zero at, e.g., 60%, then one may assume ideality and calculate the mole fraction from

$$RH^* = 100(1-x_S) \qquad (9.33)$$

where x_S is the mole fraction of solute (drug) at saturation. Ideality is not Utopian, because in most cases (due to the high molecular weights of drug in relation to that of water), x_s will be small.

To use the latter it is best to use a non-linear curve fitting program to fit segments DC in Figure 9.8.

If an anhydrate is exposed to a series of relative humidities and moisture uptake rates recorded (i.e. the initial slopes of graphs such as Figure 9.14 are measured), then plots such as shown in Figures 9.9 and 9.10 result. The discontinuity at B in Figure 9.9 occurs at the relative humidity of the hydrate, and the discontinuity at D at the relative humidity over the saturated solution. In case of multiple hydrates (e.g., some sodium phosphates have as much as four hydrate forms, 2, 5, 7, and 12) there will be as many knees in the curves as there are hydrates.

If the compound does not form hydrates, then a curve such as shown in Figure 9.10 occurs (VanCampen et al., 1980). The intersection with the x-axis corresponds to the relative humidity over a saturated solution of the compound.

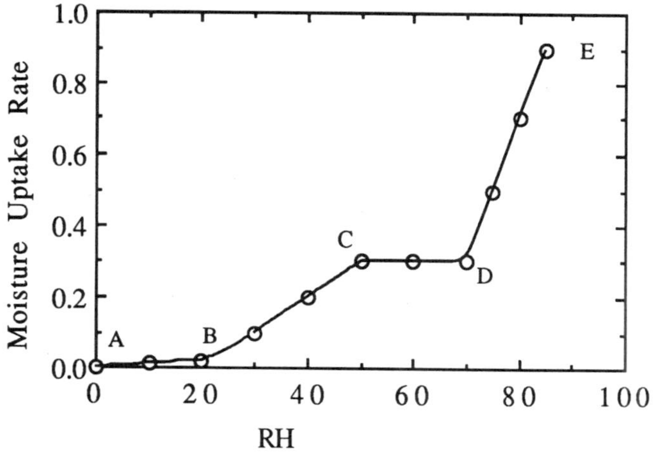

FIGURE 9.9 Moisture uptake rates of a compound which forms a hydrate.

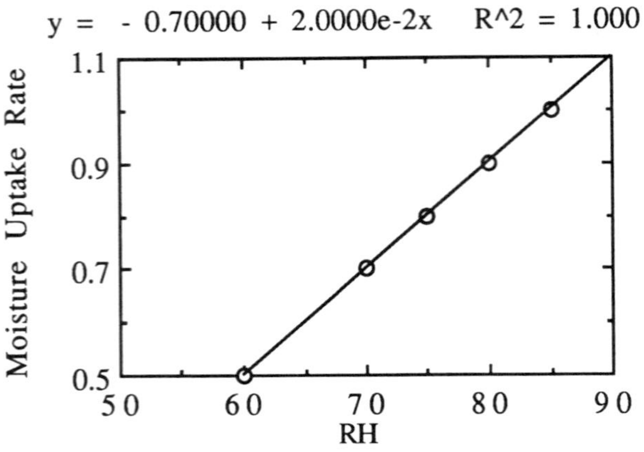

FIGURE 9.10 Moisture uptake rates of a compound which does not form a hydrate.

9.9 SOLUBILITIES FROM MOISTURE ISOTHERMS

Data such as used in Figures 9.9 and 9.10 may be utilized to obtain solubilities with relatively small amounts of sample. The ordinate value of A in Figures 9.7 and 9.8 directly gives the amount of water per g of dry solid for a saturated solution, and this can be converted to g of dry solid per 1000 g of water, i.e., to the conventional concentration unit used for solubility.

If microcalorimetric means are used in early development (when only small amounts of drug are available), and later shored up with volumetric isotherms, this calculation can, of course, not be done until the latter has been obtained.

9.10 PREFORMULATION TASKS

The task of the preformulator at the onset is an assessment of the hygroscopicity of the compound. With small amounts of solid a vapor pressure isotherm may be obtained by means of microcalorimetry. The heat involved as a function of relative humidity is measured in a thermal activity monitor equipped with a flow tube arrangement. An air stream is introduced, at first dry nitrogen, then nitrogen containing 10% moisture, and so on.

When the heat is plotted as a function of relative humidity a curve similar to the ones shown in Figures 9.6 and 9.8 will occur. The ordinate, however, is in Joules per gram of dry solid, not grams of water per gram of dry solid.

Larger amounts, when they become available, are treated similarly, and one larger sample sacrificed, e.g., exposed to 40%RH by other means than the calorimeter (a desiccator for instance). The microcalorimetric heat at 40%, can now be equated to the heat response at 40%, and the entire curve, both for the larger and for smaller samples, can be translated into conventional isotherms.

Pro forma, it should be possible to obtain the term V_m in the BET equation from this, but this term is not necessarily translatable into a surface area term. The cross-sectional area of water has been reported to be 10 or 12.5Å^2 respectively, and the question whether water closely packs on the surface is moot.

The task of assessing solubility by isotherms is an added benefit, since, in most cases, isotherms of type 9.6 and 9.7 are at hand.

If capsules or direct compression are contemplated, then isotherms of common excipients used in the company should be established. For some of these (microcrystalline cellulose, corn starch), surface areas are fairly well established and need be done only once or periodically. For others (magnesium stearate, talc, modified cornstarch), ranges should be established.

The important aspect is that no ingredient should be at a moisture content that would allow the drug substance to pick up moisture from it (the Gore-Ashwin rule). This moisture content may be deduced from comparison of the two isotherms.

NOMENCLATURE

A	Constant (at given RH) defined by Eq. 9.16
B	Constant (at given RH) defined by Eq. 9.17
D	Diameter of the solid plus adsorbed liquid (cm)
d_s	Diameter of the solid particle (cm)
d_o	initial diameter of the solid particle (cm)
h	thickness, cm, of the adsorbed layer
m	Initial weight ($\rho \pi d_o^3/6$) of solid particle
N	Number of particles per gram
P_o	Vapor pressure of water at the temperature in question
P_a	Water vapor pressure in the atmosphere at the temperature in question
P_s	Water vapor pressure over a saturated solution of the solid
ρ	True density of the solid (g/cm^3) at the temperature of study
ρ_o	True density of the water (g/cm^3) at the temperature of study
ρ^*	Composite density of the whole particle (solid (g/cm^3) + liquid layer) at the temperature of study
RH_a	($=100P_a/P_o$), relative humidity in the atmosphere at the temperature in question
RH_s	($=100P_s/P_o$), relative humidity over a saturated solution of the solid
s	Solubility (g solid/g of water) of the solid in water
t	Length of time of exposure of the solid to the humid atmosphere
w	Gram of water adsorbed by one particle at time t at the temperature of study
x_s	Volume fraction of solid dissolved in aqueous layer at saturation

REFERENCES

Brunauer, S., Emmett, P.H. and Teller, E., (1938), *J. Am. Chem. Soc.*, **60**:309.

Brunauer, S., Katro, D.L. and Wiese, C.H., (1959), *Can. J. Chem.*, **37**:714.

Carstensen, J.T., Danjo, K., Yoshioka, S. and Uchiyama, M., (1987), *J. Pharm. Sci.*, **76**:548.

VanCampen, L., Zografi, G. and Carstensen, J.T., (1980), *Int. J. Pharm.*, **5**:1.

Yoshioka, S. and Uchiyama, M., (1986a), *J. Pharm. Sci.*, **75**:92.

Yoshioka, S. and Uchiyama, M., (1986b), *J. Pharm. Sci.*, **75**:945.

CHAPTER 10

Principles of Dissolution

Dissolution is a most important property in pharmaceutics, and the dissolution characteristics of a solid drug substance must always be assessed. The writing to follow is a brief outline of dissolution principles. There are two principal types of dissolution to be discussed (Figure 10.1). One is dissolution where the surface area stays constant during the dissolution (plate dissolution), and one is where a powder dissolves in a liquid, and hence, as the substance dissolves, the surface area changes (particulate dissolution).

10.1 THE NOYES-WHITNEY EQUATION

The rate with which a solid dissolves into a dissolution medium (or biological fluid) is a function of (a) its surface area, A, and (b) its solubility and the velocity, v, with which the liquid passes by the surface of the dissolving solid.

At a given velocity, the rate may be described by the Noyes-Whitney equation (Noyes and Whitney, 1897):

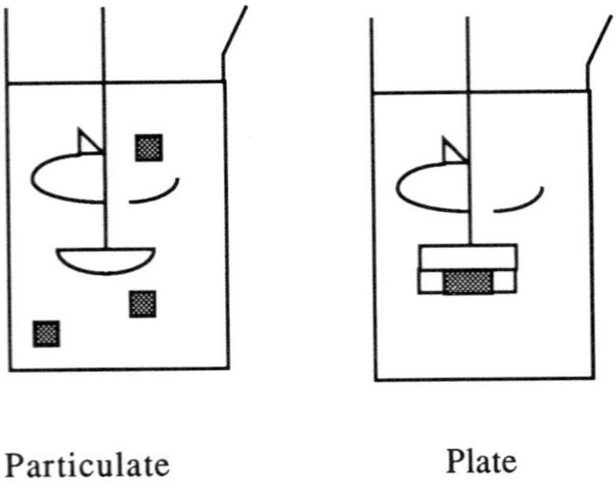

FIGURE 10.1 The two dissolution types.

$$dM/dt = -kA(S-C) \qquad (10.1)$$

where S is solubility, C is the concentration at time t and k is a so-called intrinsic dissolution rate constant. It can be seen from the equation that the dimension of k is cm/sec.

Dissolution is monitored by starting experiments of the type shown in Figure 10.1 and sampling the liquid from time to time. If the volume of dissolution liquid is L, then the amount dissolved is VL, and the amount not dissolved is M_0 - VL, then Eq. 10.1 will take the form:

$$LdC/dt = -kA(S-C) \qquad (10.2)$$

If A is constant, then Eq. 10.2 integrates to:

$$\ln[1-(C/S)] = -(kA/L)t \qquad (10.3)$$

At the onset of dissolution, and in cases where a large amount of liquid is used,

$$C \ll S \tag{10.4}$$

This is referred to as sink conditions, and in this case Eq. 10.2 simply integrates to:

$$C = [kAS/L]t \tag{10.5}$$

and this is assumed to hold if $C \ll 0.15S$.

The use of a compressed disk is referred to as a Wood's apparatus (Wood et al., 1963) when the set-up is as shown in the second drawing in Figure 10.1. In this manner it is felt that the area will stay constant (Figure 10.2).

Plate dissolution is, at best, an estimate, and probably a fairly inaccurate estimate, of dissolution. The general idea is that the surface that is exposed to the liquid (Figure 10.2A) will stay constant during the dissolution and simply recede (Figure 10.2B). However (Carstensen, 1977), it is more likely that it will become uneven during dissolution (Figure 10.2C). The other problem is that often too little compression pressure is applied to prevent the compact from being porous. This, essentially, should cause curvature in the dissolution plotted according to Eq. 10.5. Often as little as 1 ton pressure per cm^2 is applied, and only for substances with very low elastic yield value will this suffice to make a non-porous compact.

Nevertheless, the method is useful, because it provides some measure of comparison, e.g. between the k-values for different salts of a drug substance. But even in such a case direct comparison is not exacting, because the substances may have different yield values and hence give different porosities.

The best method is to repeat an experiment several times using different compression pressures until a consistent value of k is obtained. Low pressures [such as used by Chan and Grant (1989)] lead to wrong k-values and erroneous conclusions.

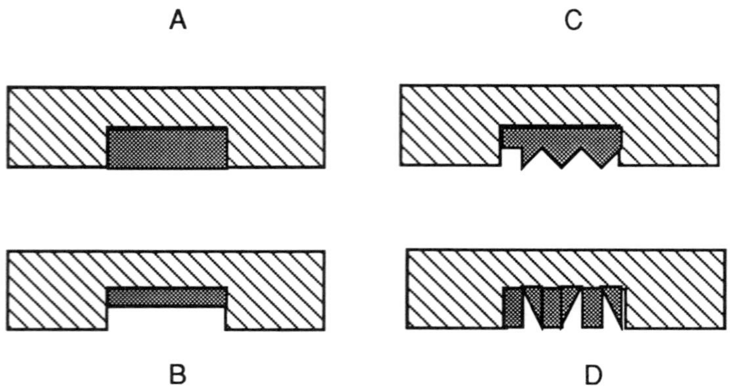

FIGURE 10.2 Types of compact formations in plate dissolution.

One reason that low pressures are preferred by many investigators is that higher pressures may lead to polymorphic transformations. *It is always advisable to remove a small amount of solid from the back of the compact and test it for morphology* (Guillory, 1992). A better (more accurate, but less precise) method will be discussed at the end of Chapter 11.

10.2 THE DISSOLUTION RATE "CONSTANT"

It should be emphasized that the value of k is not a universal constant, but rather depends on the hydrodynamics of the system (Carstensen and Dhupar, 1976). It can be shown that, to a certain approximation,

$$k = \beta v^{1/2} \tag{10.6}$$

where v is the linear velocity of the liquid passing by the surface of the dissolving substance.

More to the point is that the plate dissolution apparatuses essentially measure something different that what was intended. Levich (1962) in studying *electrodeposition* on a rotating plate (i.e. the opposite of dissolution), arrived at the following rate equation:

$$\text{Rate} = D\eta\omega^{1/2} \qquad (10.7)$$

where D is diffusion coefficient, η is kinematic viscosity of the liquid, and ω is the revolutions per time unit.

It is noted that the method does not suffer from the problems stated in Figure 10.2, since it is a deposition, not a dissolution. For many years, physical pharmacists considered that

$$K = D/h \qquad (10.8)$$

the so-called film theory, where h would be the thickness of a somewhat stagnant film about the particle. One of the problems associated with this is that when a particle starts dissolving it may have a film of a given thickness, but will this remain the same during the dissolution process? For instance, when the particle is just about to completely dissolve, could it have a film thickness of any measurable thickness? This problem will be assessed in the next chapter. There are other problems of various kinds with this assumption of Eq. 10.8, and they will not be considered in this text. What *will* be assumed is the validity of the Noyes-Whitney equation, without interpretation of what k may mean.

10.3 PARTICULATE DISSOLUTION

The plate dissolution method gives a value of k, but it would never be the same as that encountered by a powder dissolving in a liquid (Figure 10.1A). The basis for dissolution of particulates is the Hixson-Crowell equation (Hixson and Crowell, 1931).

Assume N cubical particles of side length a, dissolving isotropically under sink conditions in a dissolution medium. The mass, M, of the particles is given by:

$$M = Na^3\rho \tag{10.9}$$

and their area is:

$$A = N6a^2 \tag{10.10}$$

The Noyes-Whitney equation now takes the form:

$$dM/dt = (dM/da)(da/dt) = 3Na^2\rho\, da/dt = kN6a^2S \tag{10.11}$$

so that

$$da/dt = 2kS/\rho = K \tag{10.12}$$

Hence,

$$a = a_o - Kt \tag{10.13}$$

or

$$1 - (a/a_o) = (K/a_o)t \tag{10.14}$$

If the volume of the individual particle is denoted v, then, noting that

$$a = v^{1/3} \tag{10.15}$$

it follows that

$$1 - (v/v_o)^{1/3} = 1 - (M/M_o)^{1/3} = K't \tag{10.16}$$

where

$$K' = 2kS/(a_o\rho) \qquad (10.17)$$

Eq. 10.16 is known as the Hixson-Crowell cube root equation. There are a series of modifications to it (Carstensen and Musa, 1972; Brooke, 1973, 1974; Lai and Carstensen, 1978; Carstensen and Dali, 1998), but the reader is referred directly to those articles for further refinement of the above.

10.4 MULTIPARTICULATES

The considerations in Section 10.3 refer to particle populations where all the particles have the same size. Only few instances in pharmaceutics, however, deal with these so-called monodisperse populations. Examples are ragweed pollen. Some substances have narrow particle size distributions, e.g., corn starch, which has particles between 25 and 35 μm.

When powders have more than one particle size they are referred to as polydisperse and these will be discussed further in Chapter 11. In dissolution of monodisperse powders there will be a given point in time, t*, when all the particles completely dissolve. This time (Eq. 10.6) occurs when:

$$t^* = a_o/K \qquad (10.18)$$

In polydisperse powders (Figure 10.2) the situation is such that there will be a smallest particle (with size a_{min}) and a largest size (a_{max}). In this case the critical time is given by:

$$t^* = a_{min}/K \qquad (10.19)$$

The situation in the case of a monodisperse powder is depicted in Figure 10.3 and the situation in the case of a polydisperse sample is depicted in Figure 10.4. For the polydisperse powder

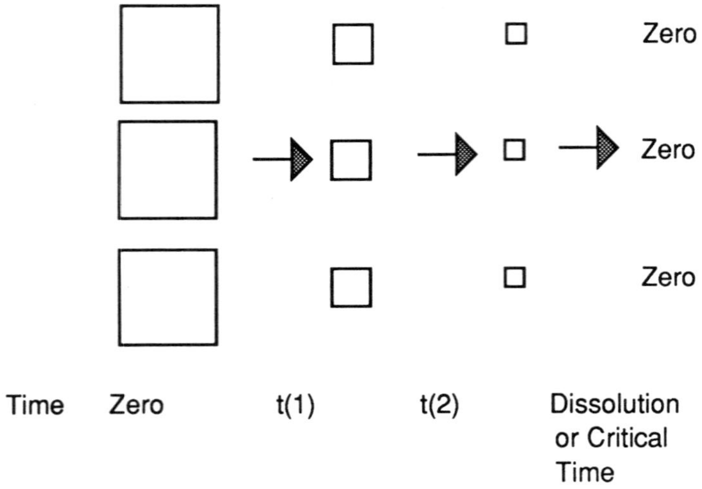

FIGURE 10.3 The situation in the case of dissolution of a monodisperse powder.

there is a critical time, when the smallest particle disappears, and a dissolution time, when the largest particle disappears.

10.5 EQUIVALENT "DIAMETERS"

It has been customary in the part of the pharmaceutical sciences where dissolution and particle size measurement is concerned to convert volume to equivalent "diameters." If a method measures the volumes of particles (electronic counters for instance), then the reported value (or the value used in dissolution work) traditionally has been the sphere which has the same volume as the particle. In fact, the Hixson-Crowell law was derived in that fashion (not with a cube, as employed as a model above). This premise, i.e., that the "size" of the particle is the side length of a cube (or the spherical equivalent diameter), is often accepted.

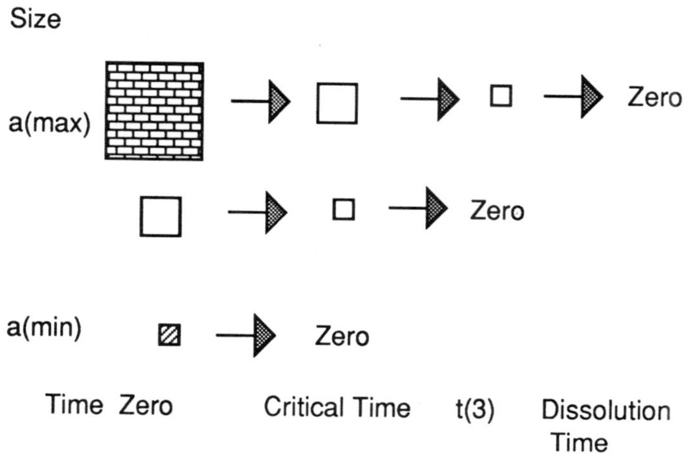

FIGURE 10.4 The situation in the case of dissolution of a polydisperse powder.

In such a case the dissolution rate constant may be determined for a solid drug substance under the correct hydrodynamic conditions (Carstensen and Dali, 1998).

If a powder is sieved into sieve fractions, then these fractions will each have a given particle size range, i.e. a minimum and a maximum size. It is often assumed that the "size" is the aperture in the sieve, b_{min}, and if this is assumed to be correct, then a dissolution rate constant may be deduced which has some meaning hydrodynamically.

The phenomenon that occurs is that seen if one plots the dissolution data of a mesh cut. Figure 10.5 shows such a dissolution profile. If it is plotted by Eq. 10.13, then the curve will have two segments, one prior to the critical time, where the number of particles remains unaltered. (In Figure 10.4 there are three particles prior to the critical time.) Then after the critical time the number of particles continuously decreases. An example of this is shown in Figure 10.6.

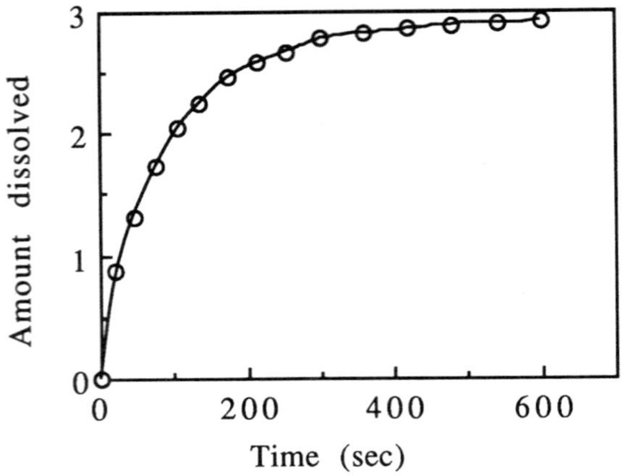

FIGURE 10.5 Dissolution of a 40/50 mesh cut of oxalic acid.

As seen in Figure 10.6, there is a nick in the curve at the point in time when the smallest particle disappears (Carstensen and Patel, 1975). If an equivalent side length of the crystals is assumed, then the value of K may be calculated from the time the nick in the curve occurs, t*, the critical time. To obtain the value of the intrinsic dissolution rate constant, k, Eq. 10.17 will have to be used.

The fallacy is the assumption of equivalent diameter, a problem which will be assessed in the next chapter. However, if consistency is maintained within dissolution data of a given compound, then the use of an equivalent dimension simply puts a constant bias on all the data collected, and the results may be used to monitor, for instance, consistency of batches produced in drug substance manufacture (synthesis).

An example of good correlation of calculated k-values (Carstensen and Dali, 1998) is shown in Table 10.1.

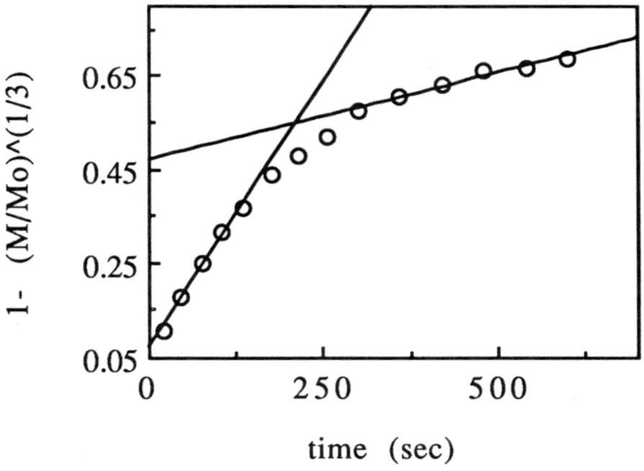

FIGURE 10.6 Dissolution of a 40/50 mesh cut of oxalic acid.

10.6 PREFORMULATION TASKS

A plate dissolution should be carried out at different applied pressures. The die size should be small when only small amounts of drug are available.

A small amount of powder should be sieved, and a microscopic particle size distribution should be carried out on each sieve fraction.

A dissolution test should be carried out on each mesh cut.

TABLE 10.1 k-values for Different Mesh Cuts of Oxalic Acid Dihydrate

Mesh fraction	k µm/sec
-30/+40	1.4 ± 0.1
-40/+50	1.5 ± 0.2
-50/+60	1.4 ± 0.4
-60/+80	1.4 ± 0.1

REFERENCES

Brooke, D., (1973), *J. Pharm. Sci.*, **62**:795.

Brooke, D., (1974), *J. Pharm. Sci.*, **63**:344.

Carstensen, J.T. and Musa, M.N., (1972), *J. Pharm. Sci*, **61**:1112.

Carstensen, J.T., (1977) *Formulation and Preparation of Dosage Forms*, ed. Polderman, J., Elsevier/North Holland, Amsterdam, p. 200.

Carstensen, J.T. and Patel, M., (1975), *J. Pharm. Sci.*, **64**:1770.

Carstensen, J.T. and Dhupar, K., (1976), *J. Pharm. Sci.*, **64**:1634.

Carstensen, J.T. and Dali, M., (1998), *Drug. Dev. Ind. Pharm.,* in Press.

Chan, H.-K and Grant, D.J.W., (1989), *Int. J. Pharmaceutics*, **57**: 117.

Chow A.H.-L., Chow, P.K.K., Zhongshan, W., and Grant, D.J.W., (1985), *Int. J. Pharm.*, **24**: 239.

Guillory, K., (1992), personal communication.

Hixson, A. and Crowell, J., (1931), *Ind. Eng. Chem.*, **23**:923.

Lai, T. Y-F. and Carstensen, J.T., (1978), *Int. J. Pharmaceutics*, **1**:33.

Levich, V.G., (1962), *Physicochemical Hydrodynamics*, Prentice-Hall, Englewood Cliffs, N.J., p. 342.

Noyes, A. and Whitney, W., (1897), *J. Am. Chem. Soc.*, **23**:698.

Wood, J.H., Catacalos, G. and Lieberman, S., (1963), *J. Pharm. Sci.*, **52**: 296.

CHAPTER 11

Micromeritics

Particle size distributions and surface areas of drug *substances* are properties of great importance both in bioavailability and in the processing of drug *products*. There should, therefore, be an effort to establish a running file of these properties as a function of lot of drug substance synthesized as development progresses. The preformulator is the first scientist to be able to assess this on the small samples he receives at the onset of the development project.

Micromeritics is the science of particle sizes, particle shapes, size distributions and surface areas. All of these are a function of how the drug substance was recrystallized and how (and if) it was milled or comminuted in some way after the polish.

As just mentioned, many pharmaceutical properties are functions of particle "size," shape, and surface characteristics. The latter two will be discussed in the following chapter. As an

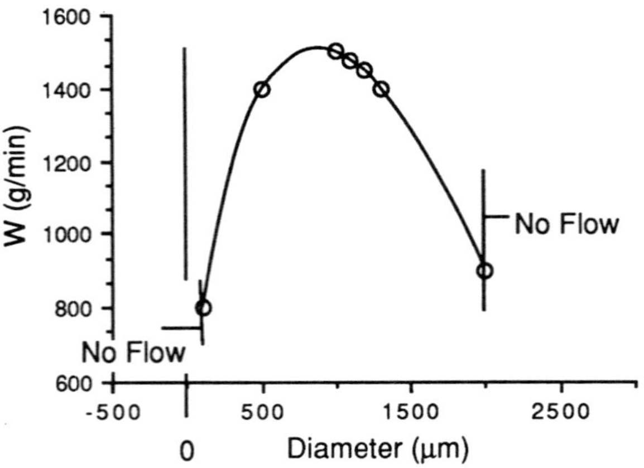

FIGURE 11.1 Flow rate as a function of "diameter," in this case the average between neighboring mesh openings.

example of the importance of the particle size, of how it affects flow has been reported often, for instance by Carr (1965), as exemplified in Figure 11.1.

The first opportunity to investigate shapes of crystals occurs in preformulation and lays the groundwork for subsequent micromeritic work. What is important, at the time of preformulation, is that properties are monitored, so that the substance, however small the sample, may be compared with subsequent lots.

The writing to follow will deal with methods that may apply to the preformulator in his quest for the maximum amount of information about a new drug substance.

There are, as mentioned, several aspects, which will be discussed in sequence: (a) particle size distributions, (b) shape factors and (c) surface areas.

11.1 MICROSCOPY

Particle size distributions of small samples may be assessed by microscopy. The best manner of doing this is by means of imaging equipment.

Table 11.1 shows microscopic data for a 40/50 sieve fraction of oxalic acid dihydrate. It is customary to count N = 200-300 particles, and to group these counts into size intervals (Dali, 1997). It is seen from the first two columns that the small dimension ranges from 202 to 436 (μm). The question then is into how many intervals, k, one now should subdivide the particles. This is given by Sturge's rule:

$$k = 1 + 3.322(\log_{10}N) \tag{11.1}$$

TABLE 11.1 Calculation of the Frequency Distribution Function from the Number Frequency for a -40/+50 Mesh Fraction of Oxalic Acid Dihydrate

Lower limit of interval (x_i)	Upper limit of interval (x_u)	Midpoint of interval (μm)	Number frequency N	F = N/N_{total}	Y = f(a) = F/26 (interval =26)
202	228	215	14	0.0558	0.00214
228	254	241	32	0.1275	0.0049
254	280	270	43	0.1713	0.00659
280	306	293	44	0.1753	0.00674
306	332	319	45	0.1793	0.00689
332	358	345	40	0.1594	0.00613
358	384	371	18	0.0717	0.00276
384	410	397	9	0.0358	0.00138
410	436	423	6	0.0239	0.00092

For N = 300 this estimates about 9 intervals, as has been done in the table.

One now calculates the *frequency*, F, i.e. in this case one divides each count by 231 (since that is the number of particles measured).

The question, then, is what type of distribution the particle sizes follow. Many programs, when data are entered into them, will show whether a distribution is normal. This is not covered here but leave suffice it to say that the particle size distributions of mesh cuts are in general normal or log normal. This is pictured in the following fashion.

The normalized ordinate values, Y = f(a), are obtained from the frequency, F, by noting that normalization (dictating unit area under the curve) requires that (Carstensen, 1996)

$$Y = F/(x_u - x_i) \tag{11.2}$$

where x_u is the upper limit and x_i the lower limit of the interval. In so doing, the data in the last column result, and these are shown in Figure 11.2.

The curve shown is the ordinates of the normal error curve applied to the data in Table 11.1, and it is seen that the data, visually, are distributed normally.

The mean dimension (e.g. the breadth of the particle) obtained in this fashion is denoted the arithmetic mean dimension, a_n, i.e.

$$a_n = \Sigma N_i a_i / \Sigma N_i \tag{11.3}$$

or

$$a_n = \Sigma f_i a_i \tag{11.4}$$

where N_i is the number of particles of dimension a_i and f_i is the fraction of particles with that dimension. The notation "a" in the

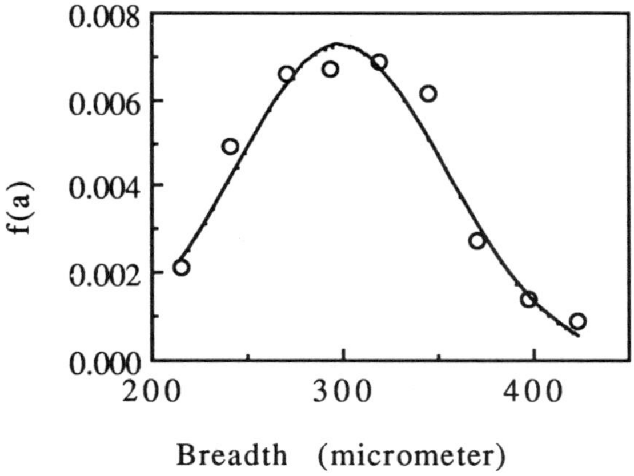

FIGURE 11.2 Distribution of the breadths of the particle for a 40/50 mesh fraction oxalic acid dihydrate.

following will be used to denote a (non-specified) dimension (length, height, breadth, diameter), whereas d will denote the diameter of an equivalent sphere (i.e. a sphere that has the same volume as the particle).

11.2 GEOMETRIC SURFACE AREAS

Geometric surface areas are defined as the area calculated from length dimensions of a solid, assuming that all surfaces are smooth (Carstensen, 1977). It is noted from Table 11.1 that it is possible to calculate the geometric surface area, if circular equivalent radii are used.

The distribution need not be normal, and in fact, often is log-normal. Such a set of data are shown in Figure 11.3 as a bar diagram.

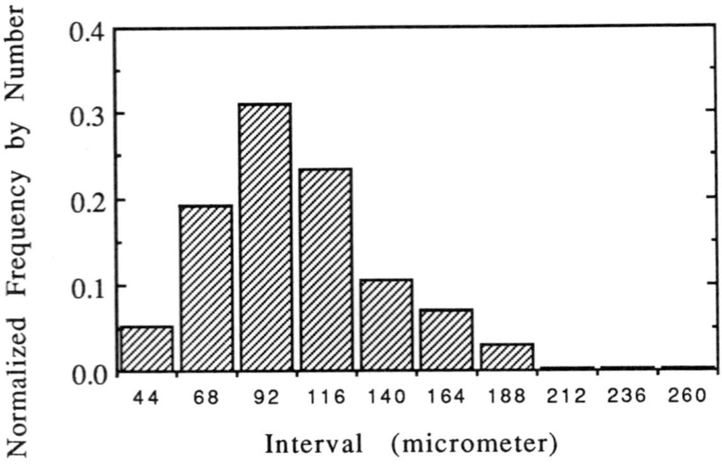

FIGURE 11.3. A bar diagram of a log-normally distributed mesh cut.

It is obvious, from visual inspection, that these data are skewed towards the higher particle sizes, and this population is actually approximately *log-normal.* If the data are plotted on probability paper as Z-value versus log of diameter, then a plot such as shown in Figure 11.4 results.

The Z-values are obtained by calculating

$$Q = (\ln d - \ln d_g)/\ln\sigma \tag{11.5}$$

where $\ln d_g$ is the average of the lnd values, and where $\ln\sigma$ is the standard deviation of the lnd values. The "area" value (Z) corresponding to Q is then found in a normal error table and plotted versus the value of lnd.

But simpler is plotting on probability paper.

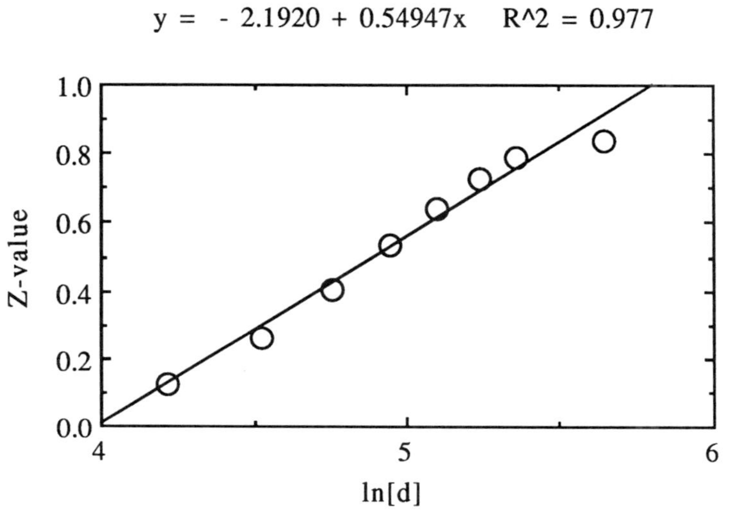

FIGURE 11.4 Data from Figure 11.3 plotted by probability means.

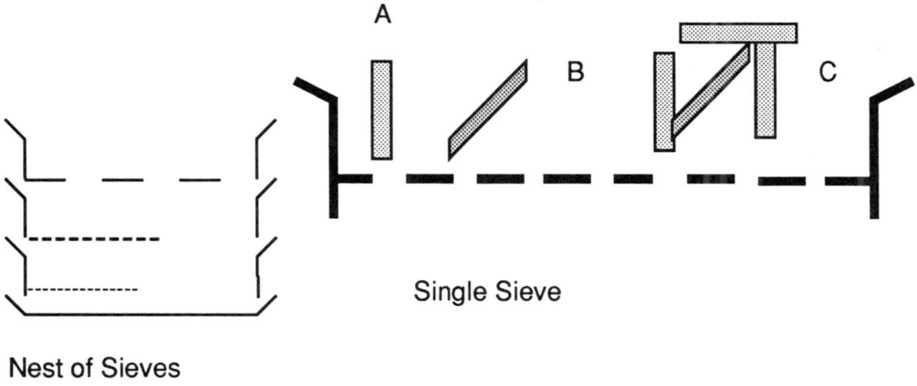

FIGURE 11.5 Principle of sieving.

TABLE 11.2 Mexh Openings

U.S. Standard Mesh	Opening (μm)
4	4760
6	3360
8	2380
10	2000
12	1680
14	1410
16	1190
18	1000
20	840
30	590
40	420
50	297
60	250
70	210
80	177
100	149
120	125
140	105
170	88
200	74
230	62
270	53
325	44
400	37

11.3 SIEVE ANALYSIS

Particle sizes are often assessed by sieve analysis. As shown in Figure 11.5, a sieve is a mesh stretched out in a containing cylinder. A nest is made of such sieves placed on top of one another, the one with the coarsest mesh on top, and so on.

The mesh number is the number of strands per running inch. The side lengths of the square opening for various mesh sizes are listed in reference works (e.g. the United States Pharmacopoeia), and an example is shown in Table 11.2.

Since the mesh opening (if the thickness of the wire were disregarded) should be inversely proportional to the opening, a graph of 1/Mesh should be linear in opening.

That this is approximately so is shown in Figure 11.6.

The method is used by placing 100 g in the top screen, and shaking the nest of sieves. As shown in the right part of Figure 11.5, the principle is that particles that are smaller than the sieve opening will pass through, and those larger will be retained.

For the time being we shall assume this to be the case.

The content of each sieve is weighed, and a table of the resulting weights is denoted a sieve analysis. The "mean" particle size is denoted the volume or weight mean diameter, i.e.

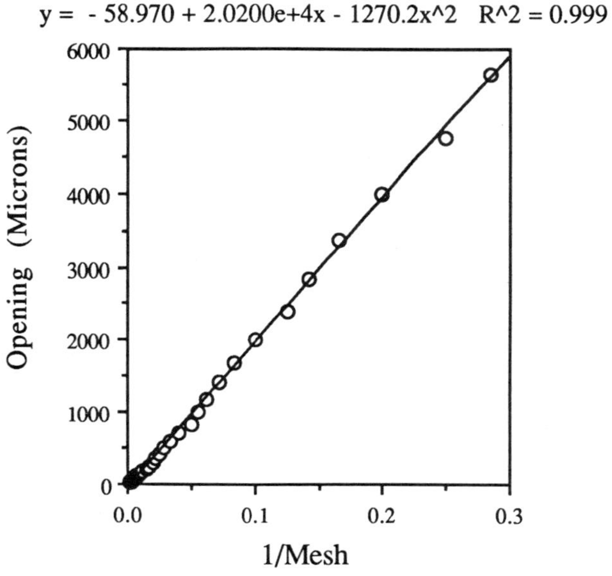

FIGURE 11.6 A plot of sieve openings versus the reciprocal of the corresponding U.S. mesh number.

$$d_{vm} = \Sigma w_i d_i / \Sigma w_i \qquad (11.6)$$

This method has the advantage over other methods of allowing a generous sample size, and hence, in that respect, gives a better picture of a large batch of powder than does more investigative and quantitative methods such as microscopy and electronic counting methods.

The general procedure for obtaining the "particle size" of a mesh cut is by knowledge of the mesh opening of the confining screens, ϕ and γ. Consider, for example, the distribution of an unsieved sample, as shown in Figure 11.7. If the sieve openings were 100 and 200 μm, then it would be common practice (Dallavalle, 1948; Parrott, 1970) to assume the "average particle size" to be 150 μm. It has been suggested that, since most particle size distributions are log-normal (Carstensen and Patel, 1975), the geometric mean, $[D_g]$, should be used, i.e., in this case $\ln[D_g] = 0.5[\ln(200)+\ln(100)]$.

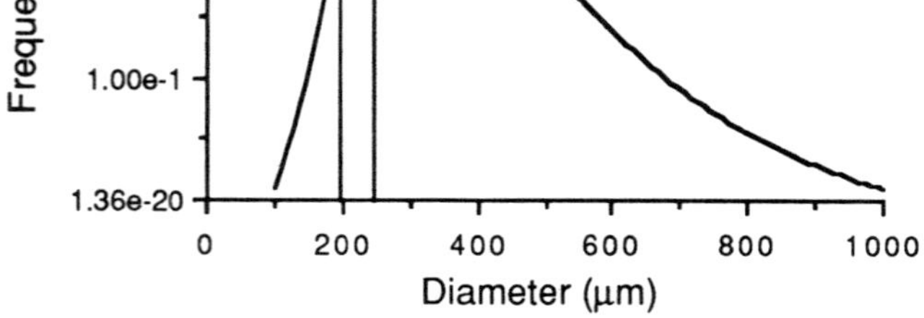

FIGURE 11.7 Particle size distribution of an unsieved, log-normally distributed powder.

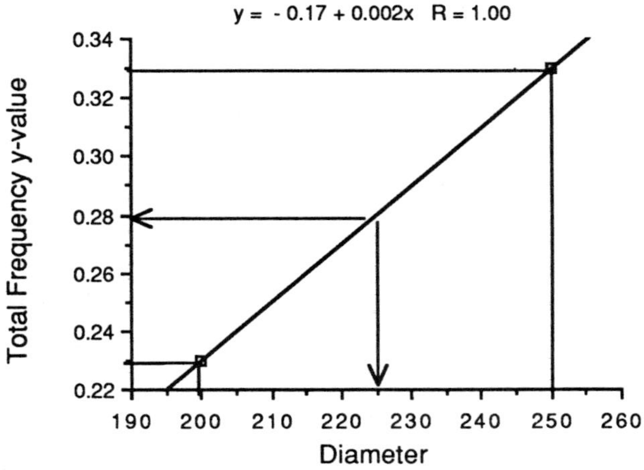

FIGURE 11.8 The expected distribution of a mesh cut between 200 and 250 μm from Figure 11.7.

As shown in Figure 11.7 it would be expected that the distribution within a mesh cut (in the figure designated as between 200 and 250 μm) would be a linear distribution. This is further illustrated in Figure 11.8.

As mentioned, however, mesh cuts may also be normally distributed. The fact that mesh cuts are not linearly distributed is, undoubtedly, due to (a) inadequacy of sieving, and (b) cohesion of smaller particles to larger ones, so that assuming the cut to be a slice out of the larger distribution is not correct.

If a particle in a powder sample has a size D, then the mass, m, of a particle will be

$$m = \rho\, v = \rho\, \alpha_v\, D^3 \tag{11.7}$$

where ρ is particle density, v is volume and the shape factor, α_v, is defined as by Dellavalle (1948), i.e.

$$v = \alpha_v D^3 \tag{11.8}$$

It should be noted that this equation may apply to any linear dimension of a particle, as long as it is chosen consistently, but the value of α_v will depend on the choice of that dimension. This dimension is what is denoted D, and shall be referred to simply as "size" in the following.

A mesh cut contains many different particle sizes, D, and these are distributed by a frequency function, f(D), which may be normalized, i.e., adjusted mathematically, so that the area under the frequency curve equals unity.

It should be pointed out that the normalized frequency function, f(D), of the particles is *logically truncated*, i.e.

$$\int_{D_{min}}^{D_{max}} f(D)\, dD = 1 \tag{11.9}$$

and as mentioned, *f(D) is not necessarily normal or log-normal, but may be of any form.*

Between D and D +dD the number of particles, n, is:

$$n = f(D)dD \tag{11.10}$$

If the mass of the entire mesh cut is denoted, M_0, then:

$$M_0 = \int_{D_{min}}^{D_{max}} n\rho\, \alpha_v\, f(D)\, D_0^3\, dD = n\rho\, \alpha_v I_3 \tag{11.11}$$

where

$$I_3 = \int_{D_{min}}^{D_{max}} f(D)\ D_o^3\ dD \tag{11.12}$$

is the third moment of the distribution. D_{max} and D_{min} are here the size of the largest and smallest particle of the powder sample and, since the object is to subject the sample to dissolution, D_o denotes the size of the particle prior to dissolution, i.e., at time zero. The dissolution experiment can be designed so that sink conditions apply during dissolution (Carstensen and Musa, 1972), so the size, D, will decrease by:

$$D = D_o - Kt \tag{11.13}$$

This, in turn, leads to the cube root equation (Hixson and Crowell, 1931):

$$1 - (M/M_o)^{1/3} = Kt \tag{11.14}$$

where K is a cube root dissolution rate constant. This holds only to a certain time, t_{crit}, at which point the smallest particle will just have dissolved, i.e.

$$t_{crit} = D_{min}/K \tag{11.15}$$

By plotting $1-(M/M_o)^{1/3}$ versus t, the curve will, at first, be a straight line then exhibit a nick, and t_{crit} may be determined in this manner. Knowing D_{min} it is, in this manner, possible to determine the value of K.

At times $t < t_{crit}$ the mass that has not dissolved from the powder in the mesh cut will be:

$$M = n\rho\ \alpha_v \int_{D_{min}}^{D_{max}} f(D)\ (D_o - kt)^3\ dD \tag{11.16}$$

After t_{crit} this no longer holds true because then n will decrease in time (since smaller particles start disappearing by completely dissolving). The cubed term in the integrand may be expanded, in which case Eq. 11.16 becomes:

$$M = (n\rho\, \alpha_v I_3) - (n\rho\, \alpha_v 3KI_2)t + (n\rho\, \alpha_v 3K^2 I_1)t^2 - (n\rho\, \alpha_v K^3 I_0)t^3 \quad (11.17)$$

where

$$I_j = \int_{D_{min}}^{D_{max}} f(D)\ (D_o)^j\, dD \quad (11.18)$$

is the jth moment of the distribution function. It is noted that $I_o = 1$ because of the normalization and that

$$n\rho\, \alpha_v I_3 = M_o \quad (11.19)$$

so that by dividing Eq. 11.17 by M_o, Eq. 11.20 results:

$$M/M_o = 1 - 3K(I_2/I_3)t + 3K^2(I_1/I_3)t^2 - (K^3/I_3)t^3 \quad (11.20)$$

By plotting M/M_o (prior to t_{crit}) as a function of t and fitting it to a third-degree polynomial in t, the coefficient to t^3 will allow calculation of I_3, because K is known. Now I_1 and I_2 may be calculated from the coefficients to t^2 and t.

Once this is known, Eq. 11.20 allows calculation of I_1 and I_2. I_1 is the mean of the population and the variance of the powder population can be calculated from:

$$\sigma^2 = I_2 - (I_1^2) \quad (11.21)$$

The value of K is given by:

TABLE 11.3. The Intrinsic Dissolution Rate Constants and Critical Times Obtained from Powder Dissolution for Mesh Cuts of Three Compounds. Data from Carstensen and Dali, 1998.

Compound	Mesh Cut	$b_{in(0)}$ (microns)	Critical time (s)	K cm/s	*k* cm/s
Oxalic acid dihydrate	-30/+40	249	220	1.1×10^{-4}	0.62×10^{-3}
	-40/+50	202	200	0.96×10^{-4}	0.56×10^{-3}
	-50/+60	148	150	0.99×10^{-4}	0.56×10^{-3}
	-60/+80	92	110	0.84×10^{-4}	0.47×10^{-3}
Salicylic acid	-40/+60	32	540	0.59×10^{-5}	2.5×10^{-3}
	-60/+100	23	420	0.51×10^{-5}	2.1×10^{-3}
	-100/+200	11	300	0.37×10^{-5}	1.6×10^{-3}
Phthalimide	-40/+60	20	2340	8.5×10^{-7}	1.4×10^{-3}
	-60/+100	14	1800	7.8×10^{-7}	1.3×10^{-3}
	-100/+200	9	1260	7.1×10^{-7}	1.2×10^{-3}

$$K = 2kS/\sigma \quad (11.22)$$

where S is the solubility and k is the intrinsic dissolution rate constant.

11.4 SURFACE AREAS

The most common micromeritic parameter used in pharmaceutics is the surface area. This is usually measured by nitrogen adsorption.

This method depends on the fact that when an evacuated solid is exposed to a nitrogen atmosphere, then the amount of nitrogen adsorbed onto the surface is a function of the nitrogen vapor pressure in the atmosphere.

In present day practice, instruments such as Quantasorb are used, and these rely on the fact that the "atmospheres" may be created by having mixtures of nitrogen and helium flow past the sample at liquid nitrogen temperature. The helium does not adsorb, and the amount of nitrogen lost from the air stream may be monitored by different means (for instance, a Pirani gauge). The amount adsorbed at a given mixture of helium and nitrogen can then be assessed by the area under the curve. This is then repeated for different nitrogen ratios (activities), and the adsorption (and desorption) isotherm are then determined.

For very small samples this may be assessed calorimetrically, as described by Pudipeddi et al. (1995).

11.5 PREFORMULATION TASKS

Surface areas by gas adsorption should be carried out.

Surface area and heats of adsorption should, if equipment is on hand, be carried out calorimetrically.

Critical diameters and dissolution rate constants should be obtained from dissolution studies of mesh cuts.

Shape factors should be determined from the dissolution studies, and attempts should be made to correlate powder dissolution with particle size distribution, dissolution rate constants and shape factors.

REFERENCES

Carr , R.L., (1965), *Chem. Eng.*, **72**: (Jan) 163; (Feb) 69.

Carstensen, J.T. and Musa, M.N., (1972), *J. Pharm. Sci.*, **61**:223.

Carstensen, J.T. and Patel, M., (1975), *J. Pharm. Sci.*, **64**:1770.

Carstensen, J.T., (1977), *Pharmaceutics of Solids and Solid Dosage Forms,* Wiley/Interscience, N.Y., p. 44.

Carstensen, J.T., (1996), *Modeling and Data Treatment in the Pharmaceutical Sciences,* Technomic, Lancaster, Pa., p. 61.

Dali, M., (1997), Powder dissolution as a tool to estimate particle size and shape parameters, Ph.D. Thesis, University of Wisconsin, Madison, WI.

Dallavalle, J.M., (1948) *Micromeritics*, Pitman Publ. Co., N.Y., p. 1313.

Hixson, A. and Crowell, J., (1931), *Ind. Eng. Chem.*, **23**:923.

Parrott, E.L., (1970), in *The Theory and Practice of Industrial Pharmacy*, ed. L. Lachman, H.A. Lieberman, and J.L. Kanig, Lea and Febiger, Philadelphia, pp. 100-119.

Pudipeddi, M., Sokoloski, T.D., Duddu, S.P. and Carstensen, J.T., (1995), *J. Pharm. Sci.*, **84**:381.

CHAPTER 12

Shape Factors

The shape and surface characteristics of crystals or particles of a compound are of great importance, since the machinability (blending properties, compression, for instance) depends a great deal on them. It is therefore important to establish a file of these characteristics as a function of lot of raw material during the development process.

The preformulator is the first scientist to assess the shape of particles of a new compound, and to follow the possible changes in this as development (and synthesis) progresses. This may be done manually and qualitatively in microscopic determinations and by other means, as shall be described below.

There have been many attempts in the literature to quantitate shape, and a couple of these will be dealt with below.

12.1 GEOMETRIC SURFACE AREA AND RUGOSITY

The geometric surface area referred to in Chapter 11 (either obtained by microscopy or by sieve analysis) assumes that all particles are spherical and smooth, and it is the smallest surface area perceivable. The real surface area will be larger because (a) the surfaces are not smooth and (b) the particles are not spherical. Frequently (Carstensen, 1980) the *rugosity*, ß, of a particle surface is obtained by

$$ß = A_{BET}/A_g \tag{12.1}$$

where A_{BET} is measured surface area (e.g. by nitrogen adsorption) and where A_g is the geometric surface area (e.g., measured by microscopy or by sieve analysis).

12.2 MICROSCOPICALLY DETERMINED SHAPE FACTORS

It is possible, as mentioned in the previous chapter, to obtain particle size distributions microscopically. If a number of particles from a sieve fraction, N, is counted, and the total weight, W, of these is determined, then, denoting by b, λ, and h the breadth, length, and height of the particle:

$$W = \rho \Sigma b \lambda h \tag{12.2}$$

Microscopically, the ratio λ/b is measured, for each particle and this is denoted q:

$$\lambda = qb \tag{12.3}$$

The smallest dimension, the height, is not directly measurable (since, if for instance the particle is box-shaped it will lie on its side, hence hiding the smallest dimension). If it is *assumed* that h is proportional to b:

$$h = \beta b \tag{12.4}$$

then Eq. 12.2 becomes:

$$W/\rho = \alpha_v \Sigma b^3 \tag{12.5}$$

where α_v is denoted the *volumetric shape factor*, given by:

$$\alpha_v = \beta q \tag{12.6}$$

Since all quantities except for α_v in Eq. 12.5 are known, this may be calculated, and then, from Eq. 12.6, ß may be calculated.

If, as is often assumed, h = b (Lu et al., 1995)

$$\alpha_v = q \qquad (h=b) \tag{12.7}$$

then the shape factor is directly and more easily obtainable microscopically.

It shall be seen in the following chapter that there are other ways of obtaining α_v. This becomes an important factor in multiparticulate dissolution.

12.3 DALLAVALLE SHAPE FACTORS

In this method (Dallavalle, 1948), spherical equivalent diameters are employed. The volume mean diameter is given by:

$$d_v^3 = \{\Sigma n_i d^3\}/N \tag{12.8}$$

The surface mean diameter is given by:

$$d_s^2 = \{\Sigma n_i d^2\}/N \tag{12.9}$$

where n_i is the number of particles with diameter d_i and N is the total number of particles. The volumetric shape factor, α_v, derives from this and is given by:

$$\text{Volume V} = \alpha_v d_v{}^3 \tag{12.10}$$

and similarly the surface shape factor, α_s, is defined by means of:

$$\text{Surface Area, A} = \alpha_s d_s{}^2 \tag{12.11}$$

Frequently a *general shape factor* denoted Γ is defined as:

$$\text{Surface Area} = \Gamma(\text{Volume})^{2/3} \tag{12.12}$$

and it is seen from this that the general shape factor is:

$$\Gamma = [\alpha_v]^{2/3}/\alpha_s \tag{12.13}$$

The general shape factor is often thought to be identical to the rugosity, ß, but this would only be true for smooth but not spherical particles (and of course for smooth spherical particles as well).

The true density, ρ (g/cm^3), of a compound is usually obtained in preformulation, and it follows that for one gram:

$$\sum N f_i \alpha_v d_v{}^3 \rho = 1 \tag{12.14}$$

f_i is the fraction of particles of size d_i, i.e., n_i/N, where n_i is the number of particles of that size (or in an interval spanning that size). For fairly monodisperse powders Eq. 12.14 simplifies to:

$$N \alpha_v d^3 \rho = 1 \tag{12.15}$$

or:

$$\ln[N] = -\ln[\alpha_v \rho] - 3\ln[d_v] \quad (12.16)$$

d_g is of course not identical to d_s or d_v, but the Hatch-Choate equations give the connection between the two, *assuming the distribution is log-normal*:

$$\ln[d_s] = \ln[d_g] + \ln^2[\sigma] \quad (12.17)$$

$$\ln[d_v] = \ln[d_g] + 1.5 \ln^2[\sigma] \quad (12.18)$$

N can be obtained approximately by counting of particles in a small, weighed powder sample.[6] If N and ρ are known, and if the value of d_v is known (Eq. 12.18), then α_v can be obtained from combining Eqs. 12.16-12.18:

$$\ln[A] = -\ln[\rho\alpha_v] + \ln[\alpha_s] - \ln[d_g] - 2.5 \ln^2[\sigma] \quad (12.19)$$

where A is the surface area. The only remaining unknown quantity in Eq. 12.19 is α_s, and the two α-values now constitute the "shape factor."

The assumptions made in the method are severe, and the method never gained great popularity.

12.4 THE RIDGWAY-RUPP EXPERIMENTS

Aside from microscopic methods only one attempt of experimentally determining shape factors, that of Ridgway and Rupp (1969), has been published in the early pharmaceutical literature.

---

6 This is fairly simple for larger particle sizes. For small particles, a micro-weighed sample is distributed on the tacky end of tape, and counting done by area, and, e.g., three counts prorated to the fully covered surface.

Ridgway and Rupp used a diamond sorter and separated sieve cuts of powders into different shapes and obtained flow rate, apparent density and repose angles. They determined that the more spherical the particle was, the better it was for pharmaceutical manufacturing.

12.5 SHAPE FACTORS IN GENERAL

There is a question regarding where a shape factor ends and some surface factor starts. It would seem logical that shape would have to be described mathematically by a *limited* set of parameters, e.g. 10, whereas roughness would require more. This could be, for instance, 40 lengths and 40 angles.

As mentioned, the general shape factor, Γ (Ridgway and Rupp, 1969; Short, Sharkey and Rhodes, 1972; Patel and Carstensen, 1975; Pothisiri and Carstensen, 1973; Lai and Carstensen, 1978) is defined as:

$$a = \Gamma v^{2/3} \tag{12.20}$$

where a is the surface area of a particle and v is its volume.

Shapes in which the shape factor is independent of dimensions are called isometric. Examples are:

Sphere	$\Gamma = 4.836$	(12.21)
Cube:	$\Gamma = 6$	(12.22)
Isometric cylinder[7]	$\Gamma = 5.536$	(12.23)

The particle in Figure 12.1A is non-isometric. If its base angle were $\pi/4$, then it would be a box with depth a, and if the other sides are denoted b and c, then the volume is:

$$v = (a/4)(c^2-b^2) \tag{12.24}$$

[7] A circle-based, right cylinder, the height of which equals the diameter of the base.

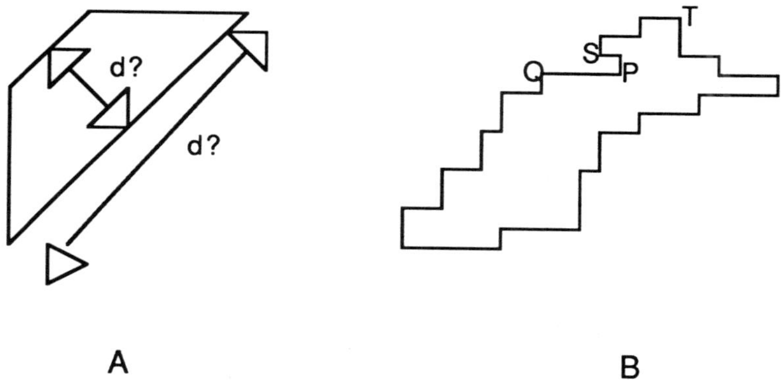

FIGURE 12.1 Distinction between "shape" and "surface roughness."

Its surface area is

$$a = h\{[(c^2-b^2)/2] + 2(2)^2(b-c) + b + c\} \tag{12.25}$$

so that even a particle of simple shape has complex shape factors, which are not independent of size (a, b and c).

It would be difficult to assign conventional mean diameters, since, as alluded to earlier, one is never certain of what the "diameter" is (Figure 12.1A and B).

12.6 DIMENSIONAL MICROSCOPIC SHAPE FACTORS

The question arises if there is a correlation between two dimensional and three-dimensional shape factors. Using the Lai (Lai and Carstensen, 1978) model for "long" cylinders, the following may be stated. The cylinder is shown in three dimensions as well as in two dimensions (as viewed under a microscope).

In three dimensions the following holds:

Volume: $V = \pi D^2 L/4$ (12.26)

Area: $A_{iii} = 2\pi DL + \pi D^2/2$ (12.27)

so the three-dimensional shape factor is

$$G_{iii} = [\pi DL + \pi D^2/2]/[(\pi D^2 L/4)^{2/3}] =$$

$$(\pi(L/D) + \pi/2)/[\{(L/D)\pi/4\}^{2/3}] \quad (12.28)$$

In two dimensions the following holds

Area: $A_{ii} = DL$ (12.29)

Perimeter $P = 2(D+L)$ (12.30)

so the two-dimensional shape factor is:

$$G_{ii} = 2(D+L)/[(DL)^{1/2}] =$$
$$2[1+(L/D)]/[(L/D)^{1/2}] \quad (12.31)$$

Calculations of each of these as a function of the length/diameter ratio are carried out by means of the program in BASIC shown in Table 12.1.

These are graphed in Figure 12.2, and it is seen that as long as a particle can be characterized by length to breadth, where two of the dimensions are fairly equal, then there is a correlation between two and three dimensional shape factors. The results are shown in Table 12.2.

TABLE 12.1 Program for Shape Factors

```
INPUT "Smallest Ratio=";Y1
INPUT "Largest Ratio=";Y2
INPUT "No of Steps-=";N
Z - (Y2-Y1)/N
REM X = L/D
FOR X = Y1 TO Y2 STEP Z
G1 = (3.14*X)+(3.14/2)
G2 = ((3.14*X/4))^(.6667)
G3 = G1/G2
REM G3 is 3 dimensional shape factor
S1 = X^(.5)
S2 = 2*(1+X)
S3 = S2/S1
REM S3 is 2 dimensional shape factor
PRINT X, G3, S3
NEXT X
```

12.7 FRACTAL GEOMETRY

It would be convenient to have a limited set of numbers to characterize the shape of a particle. For this purpose, Mandelbrot (1983) introduced the concept of fractal geometry. The general principle is the following: If the length of a contour (Figure 12.1B) were measured, then the length, L, of the perimeter of the polygon would depend on the "length of the yard stick," the so-called scaling length, λ. If this had the length AB in Figure 12.3, then the total length, AF, would be larger than if the scaling would be AC (Richardson, 1961).

It appears, in general, that the length, L, of a perimeter is a function of the scaling length, λ, by the relation:

$$\log[L] = -H\log[\lambda] + \beta \tag{12.32}$$

TABLE 12.2 Results from the Program in Table 12.1

Length/Diameter Ratio	3 Dimensional Shape Factor	2 Dimensional Shape Factor
1	5.535	4.00
2	5.811	4.243
3	6.209	4.619
4	6.589	5
5	6.940	5.367
6	7.263	5.715
7	7.562	6.047
8	7.840	6.364
9	8.101	6.667
10	8.347	6.958

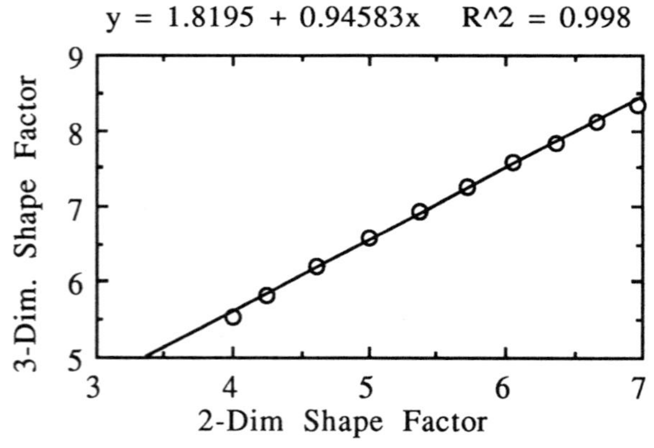

FIGURE 12.2 Correlation between three- and two-dimensional shape factors.

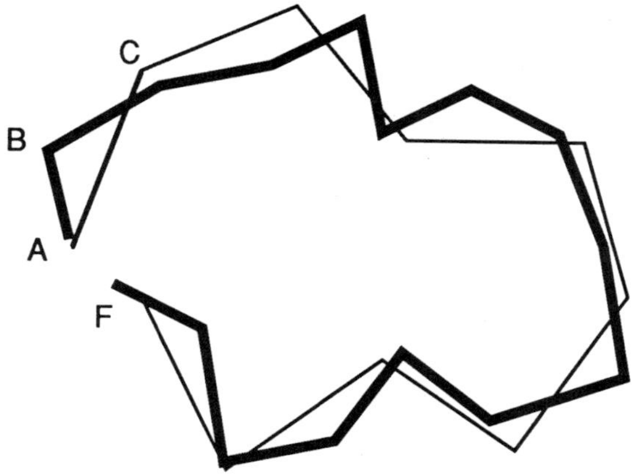

FIGURE 12.3 Fractal dimensions of a contour. The curves and concepts are approximate and are shown for definition purposes only. The point where a curve becomes a plane filling curve (e.g., a Peano curve) is complex and beyond this writing.

ß is a constant, and H is related to the fractal *dimension*, D, by the relation

$$H = 1\text{-}D \tag{12.33}$$

When plotted in this fashion, these graphs are known as Richardson plots (Figure 12.4). They are mostly plotted simply linearly, as shown in the lower line in Figure 12.4 (in which case the fractal dimension would be 1.14), but they are often bimodal showing one structure (as in the top of Figure 12.4) with a fractal dimension of 1.06 and a finer structure of dimension 1.17.

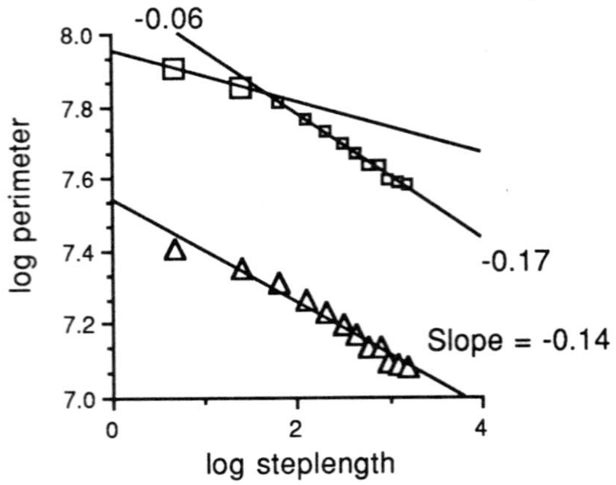

FIGURE 12.4 Example of L, the logarithm of the perimeter, as a function of the logarithm of the step length. All the data are shown in simple regression in the lower curve. The upper curve is bimodal presentation, where 0.5 has been added to the points for graphical clarity.

The interpretation of fractal dimensions is illustrated in Figure 12.5. A line, a circle or a square has one dimension. In a Eucledian sense all lines have one dimension, but if the line gets "squiggled" as shown in the shape to the left in Figure 12.5 it starts to "fill out" the plane and approaches the dimension of a plane, viz. 2.

Rather than using the approach in Figure 12.3, a mosaic approach as shown in Figure 12.6 may be used. With imaging microscopy a grid is superimposed on the shape, and a program counts the number of times, q, the contour cuts a "square" of the grid. The perimeter length is then

$$L_i = q\lambda_i \tag{12.34}$$

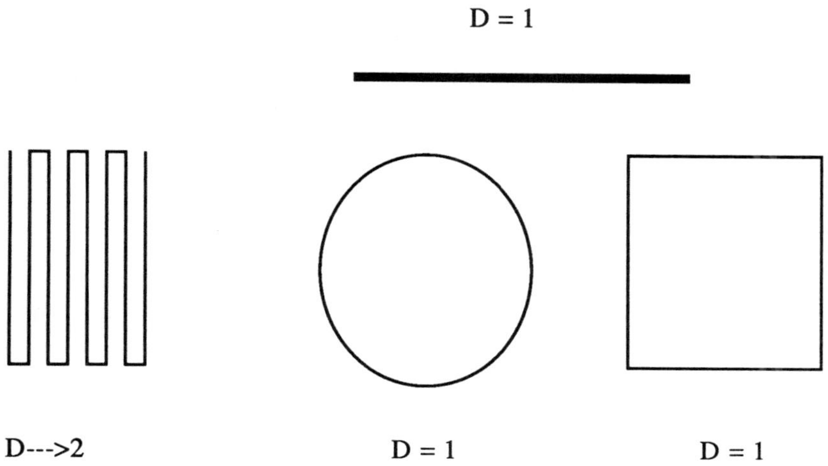

FIGURE 12.5 Examples of fractal dimensions.

where λ_i now is the side length of the square. Hence

$$\log[Li] = \log[q] + \log[\lambda_i] \qquad (12.35)$$

But this in turn equals

$$\log[L_i] = -H\log[\lambda_i] + \beta \qquad (12.36)$$

$$\log[q] = (-1-H)\log[\lambda_i] - \beta \qquad (12.37)$$

The above considerations have been based on microscopy and two-dimensional analysis. If the "probe" were a gas molecule, and the area of the solid were determined by gas adsorption using different gases, then the log of the surface area may be plotted versus the log of the diameter squared (the

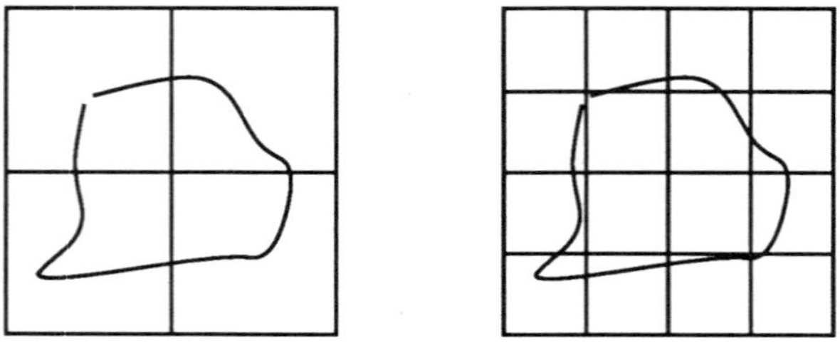

FIGURE 12.6 Principle of square grids. In the first graph, the side length is a/2 and the number of squares "cut" is four in the second, the side length is a/4 and the number of squares "cut" is 10.

surface coverage of the adsorbate) of the gas molecule, leading to fractal plots as shown in Figure 12.7.

Conversely, adsorption of one particular gas on different sieve fractions changes the "perimeter" (in this case the area), but keeps the scaling factor (in this case nitrogen gas) constant, and hence leads to a fractal graph as well.

Figure 12.7 is constructed from published data (Avnir et al., 1983) and plotted, the surface area (A) obtained by BET (nitrogen) of different sieve fractions of colloidal silica. In this case the applicable equation is:

$$\log[A] = (D-3)\log[d] + \text{constant} \tag{12.38}$$

12.8 FRACTAL ANALYSIS AND PHARMACEUTICAL PARAMETERS

Thibert et al. (1988) obtained fractal dimensions of several materials and listed several properties as a function of their respective roughness factor, H (equal to 1-D). The percentage a bed of powder can decrease in apparent volume (the compression ratio) is not its true density (since it is porous) and it depends on particle shape and size. It is therefore possible to plot them on a single graph, and a plot of the compression ratio is shown in Figure 12.8.

Other attempts to describe the surface roughness of particles have been made (Franchini and Carstensen, 1993). The data shown in the following were obtained by producing a starch-lactose granulation in a 5 kg planetary mixer and sieving it.

Cascaded apparent densities into various diameter graduates were determined and the powder was then tapped to obtain the tap density, ρ" (g/cm^3). The ratio, C, of the two figures, the so-called compression ratio, is:

$$C = \rho' / \rho'' \tag{12.39}$$

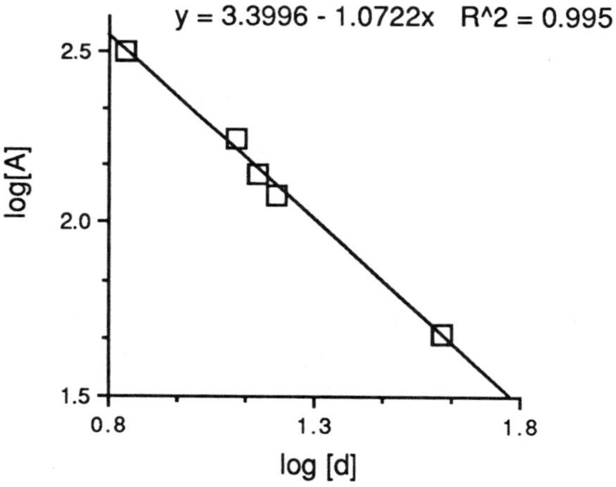

FIGURE 12.7 Fractal plot of carbon black. Amount of adsorbate, n (mmol/g) in monolayer as a function of cross-sectional area (Å^2) of adsorbing molecule.

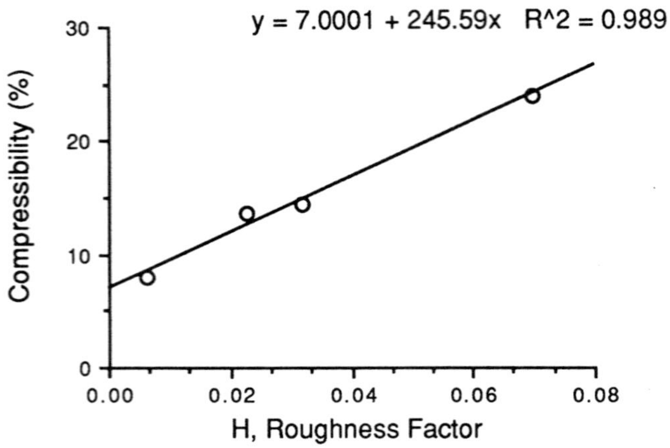

FIGURE 12.8. Compressibility ratios versus roughness factor. Plot constructed based on tabular figures published by Thibert et al. (1988).

The results failed to show significant influences of graduate diameter on compression ratio, but there were significant effects of particle diameter (mesh cut) on the value of C. Figure 12.9 shows the effect of sieve diameter on compression ratio, and Figure 12.10 shows the logarithmic transformation of this.

The relationship must be a surface phenomenon and may serve as an empirical method for describing surface roughness and particle shape.

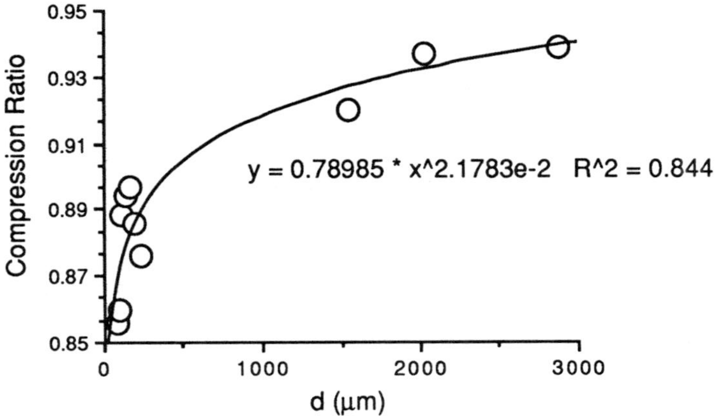

FIGURE 12.9 Compression ratio as a function of sieve opening.

12.9 PREFORMULATION TASKS

If imaging equipment is available, fractal dimensions of all batches in the clinical program should be determined.

The shape factor should be determined by the Dallavalle method by weighing and counting a small number (e.g., 300) particles, and determining the breadth/height ratio.

Two-dimensional shape factors should be determined microscopically and translated into three-dimensional shape factors by the correlation in Figure 12.2.

Adsorption experiments with carbon dioxide aside from the usual nitrogen adsorption should be carried out, and a line drawn similar to the line in Figure 12.7 to obtain the fractal dimension from the slope.

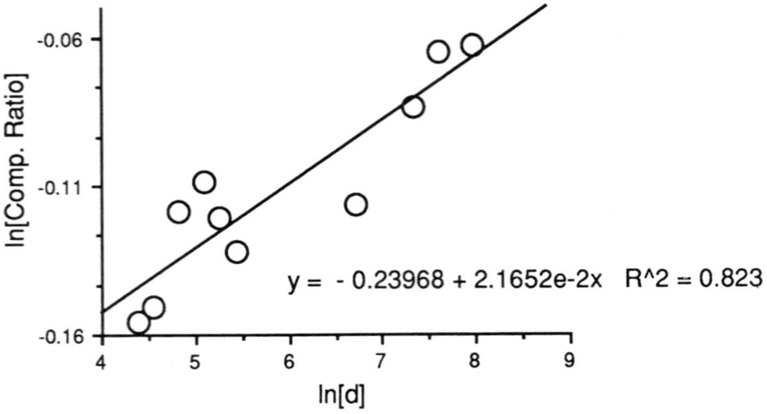

FIGURE 12.10 Logarithmic presentation of the plot in Fig. 12.9.

REFERENCES

Avnir, D, Farin D. and Pfeifer, P., (1983), *J. Phys. Chem.*, **97**: 3566.

Carstensen, J.T. and VanScoik, K., (1990), *Pharm. Res.* **25**: 1278.

Carstensen, J.T., (1980), *Solid Pharmaceutics*, Academic Press, p. 43.

Dallavalle, J.M., (1948) *Micromeritics*, Pitman Publishing Corp., N.Y., p.50.

Franchini, M. and Carstensen, J.T. (1993), *Drug Dev. Ind. Pharm.*, ***19***:85,

Lai, T.Y-F. and Carstensen, J.T., (1978), *Int. J. Pharmaceutics*, **1**: 33.

Lu, A.T.K., Frisella, M.E. and Johnson, K.C., (1995), *Pharm. Research*, **10**:1308.

Mandelbrot, B., (1983) *The Fractal Geometry of Nature*, W.H. Freeman and Company, N.Y., N.Y.

Patel, M. and Carstensen, J.T., (1975), *J. Pharm. Sci.*, **64**:1651.

Pothisiri, P. and Carstensen, J.T., (1973), *J. Pharm. Sci.*, **62**: 1468.

Ramadan, M.A and Tawashi, R., (1990), *J. Pharm. Sci.*, **79**: 929.

Richardson, L. F., (1961), *General Systems Yearbook*, **6**:139.

Ridgway, K and Rupp, R., (1969), *J. Pharm. Parmac.* **21**:30S.

Short, M.P., Sharkey, P. and Rhodes, C.T., (1972), *J. Pharm. Sci.*, **61**:1733.

Thibert, R., Akbarieh, M. and Tawashi, R., (1988), *J. Pharm. Sci.*, **77**:724.

CHAPTER 13

Physical Testing

It has been mentioned that aside from the preformulator being involved with testing and parameter evaluation before the first clinical batches are made, s/he is also involved with the initial evaluation testing of phase I, phase II and to some extent phase III material. It is not always that there is preformulation involvement, but the experiences gathered from the first clinical batches are always important, and a routine gathering of information as the project advances is advocated.

In this chapter, a short outline of the manufacturing procedure for each solid dosage form is presented, followed by the task of the preformulator for dosage forms of the particular type.

13.1 PHARMACEUTICAL EQUIPMENT FOR SOLID DOSAGE FORMS

The solid dosage forms are powders, tablets and capsules. They are usually the least expensive to produce, for example, because in shipping, the weight of a solid dosage form is considerably less than that of a liquid dosage form. They also are usually more stable than the liquid dosage forms. By far the largest percentage of dosage forms sold in the United States are solid dosage forms.

13.2 HARD SHELL CAPSULES

These are encapsulated on hard shell capsule machines. The simplest of these (and the oldest) is a two ring capsule machine. The powder is placed in a movable hopper. The capsules are in a stationary hopper. When the machine is turned on, capsules will fall through a chute, align themselves, and drop into a ring consisting of two halves. By means of vacuum the body of the capsules are sucked into the lower half of the ring, and the caps stay in the top half of the ring.

The rings are then separated, the lower part placed on a turntable over which the movable ring is placed. The table turns and powder for the fill is placed in a hopper, which is drawn over the ring as it rotates. As the ring turns, the powder flows into the body of the capsules. At the end of a rotation of the turntable the machine is stopped, the hopper is moved back, and the two rings brought together again. By means of a pegboard, the capsules are then (a) pushed together (i.e. the body is re-entered into the cap) and (b) ejected.

The method is fairly slow, and more rapid methods are used in production nowadays. However, the two ring machine is quite adequate and practical for product development of initial clinical batches which often use quite small amounts of raw material. It is noted that flow is important, so that *non-cohesive* powders are better than cohesive powders.

More modern machines (Häffliger-Karg and Zannazi) rely on a dosator. This is a cylinder with a movable piston-like base to which vacuum can be applied. The powder is sucked up into or tamped into the cavity. The amount of material filled into this cavity (the fill weight) is a function of the particle size distribution and compressibility of the powder, and of the vacuum applied. Hence, from a formulation point of view it differs from the considerations that must be given to the three ring capsules. The powder must also form a *cohesive* plug, which is not necessary in the two ring machine. The HK and Z machines have a special arrangement for separating the two capsule shells, and the dosator delivers the plug to the body of the capsule, and the top is then applied.

In recent years dosage forms have been attempted where there is a *liquid fill into the capsule (e.g., PEG bases)*. This (as well as what will be said about soft shell capsules later) is the reason that the preformulator is interested in *stability and solubility of new drug substances in anhydrous solvents*. The capsules just described go under the name of "liquid filled caplets" in the over-the-counter market in the U.S.

A formula for a hard shell capsule product may be as simple as the following three ingredients:

Drug substance
Filler (e.g., lactose hydrate)
Lubricant (e.g., magnesium stearate) 0.5-1%

The lubricant is added to facilitate the movement back and forth over the fill ring. The amount used is rarely over 1%, since large amounts of lubricant interfere with dissolution.

In a simplified way of looking at formulation, the first thing to be established is: what should the fill weight be? If 200 mg of drug is to be filled into the capsule, but if this only fills up the capsule say 2/3 of the way, then the remainder must be filled up with filler. This will be discussed further in Section 13.14. *From a preformulator's point of view, it is therefore important to have some feel for the ratio of drug to excipient.* This, of course, is now known in phase I, but some feel for it is obviously present from the pharmacological screen, so that one knows if the drug is potent (i.e., low ratio of drug to excipient), or vice versa.

From a manufacturing point of view, two batches are never going to have the same apparent density, so that under the same conditions, the fill weight could be different. In a two ring machine the capsule machine, however, can be run at different speeds. A higher speed gives less of a chance for powder to flow into the capsule, since the contact time between the capsule and the hopper is shorter. Hence, the machine speed can be used to achieve the correct fill weight.

With capsules based on the dosator principle, the fill weight can be adjusted directly with the dosator.

Apparent density of mixtures is, however, a point with which the preformulator should be concerned, and monitoring this quantity during the formulation stage is either the responsibility of the formulator or the preformulator.

There could, for instance have been five batches of raw material used in the clinical trial at NDA time. If the drug is potent, then the apparent density (hence the fill weight) is only a function of the excipient, but if there is a lot of drug in the dosage form, then properties of the drug are responsible for the value and changes in the apparent density of the mixture.

This relates to particle size, particle shape, and fractal dimension.

13.2.1 HARD SHELL CAPSULES - HYGROSCOPICITY

Hard gelatin capsules have a water content of 12-16%. If the equilibrium water vapor pressure of the capsule powder is considerably lower than that of the gelatin, then so much of the water can be drawn out of the shell that (a) the shell becomes brittle, and (b) the moisture added to the powder granulates it completely so it becomes a hard plug. In such a case the dissolution of the drug will be a square root law (treated separately below).

13.3 DISSOLUTION OF CAPSULES

For solid dosage forms (and suspensions) to be bioavailable, it is in general necessary that the drug dissolve readily from them. To this end there are three USP dissolution methods. In Europe there is a fourth (a flow-through method) which is also used often. The USP methods are:

(a) Method I: basket method
(b) Method II: paddle method
(c) Method III: reciprocating method

The principle of a disintegration apparatus is shown in Figure 13.1 and the apparatus for USP dissolution test I is shown in Figure 13.2. Tablets are placed in the rotating screen, and samples are taken from the liquid, from time to time, and assayed.

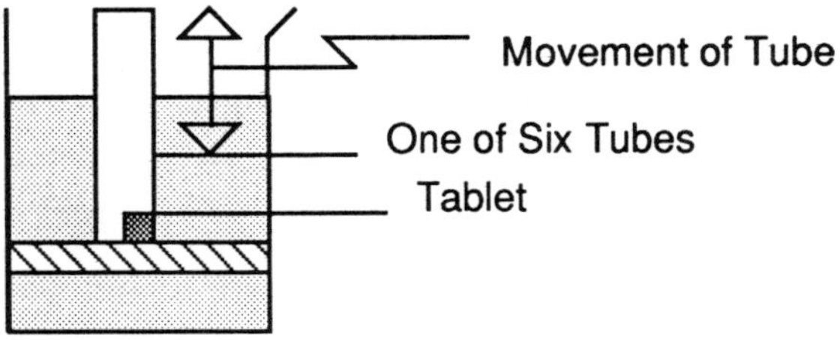

FIGURE 13.1 Principle of a disintegration apparatus.

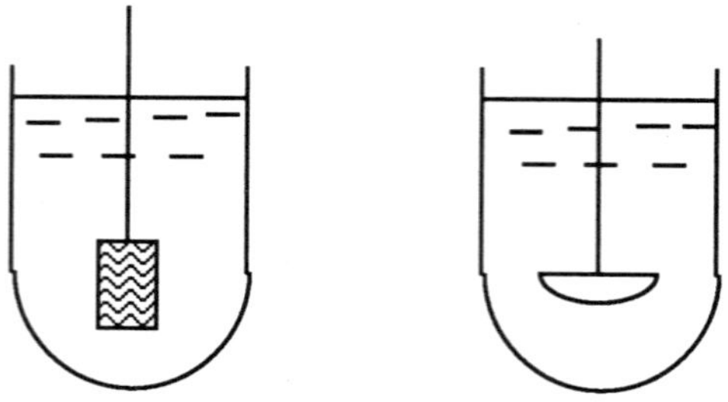

Basket Method — Method I

Paddle Method — Method II

FIGURE 13.2 USP dissolution apparatuses I and II.

Capsules when tested by USP Method II (Paddle) are usually weighted down with a metal coil, so they will sink. Upon storage, the powder, if it is sufficiently hygroscopic, may pull out moisture from the capsule shell, and "granulate," i.e. agglomerate. This may slow down dissolution (i.e., decrease, for instance, the amount dissolved after 30 minutes. Most quality control specifications are "one-point" tests, e.g., the USP often specifies that 75% must dissolve after 30 minutes). In Figure 13.3 (dissolution of a 0.5 gram tablet) the length of time for 75% (0.375 grams) to dissolve has increased from 2 to 8 minutes in 1 year's storage, and it is shown, in this case, that after 2 years the length of time required to dissolve 75% of the drug has increased to about 30 minutes.

13.4 ROUTINE TASKS WITH HARD SHELL CAPSULES

Regarding compatibilities, the most important factors in considering hard shell capsules are:

(a) the moisture isotherm (hygroscopicity) of the drug

(b) the presence of groupings that contain aldehydes or ketones, since they promote gelatin cross-linking.

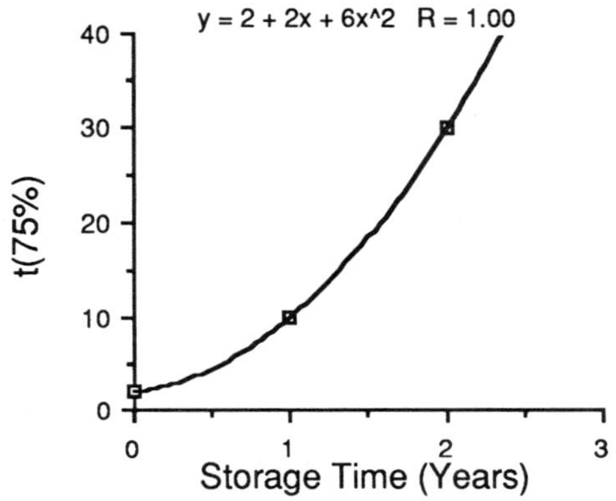

FIGURE 13.3 Increase in dissolution initially and after two years' storage at 25°C.

The latter will be discussed further under Section 13.5 (Soft Shell Capsules) and the former is discussed, on theoretical grounds, in Section 13.17.

13.5 SOFT SHELL CAPSULES

In soft shell capsules two sheets of gelatin are formed into pockets and a liquid dosed into the cavity just prior to sealing. The gelatin films are a gel (solution) of water, gelatin and glycerin. After the (wet) capsule has been formed, it is dried to a given moisture content. This cannot be too low, since then the capsule would be brittle, or too high, since then it would distort in storage. The glycerin content is also critical, since too small a content would also contribute to brittleness, and too large a content to distortability.

The fill is usually a *triglyceride* (oil) or *polyethylene glycol* (PEG, Carbowax). The former is completely immiscible with the shell, the latter presents a problem in that the glycerin is soluble in it, and to compensate for this, an equilibrium amount of glycerin is usually added to the fill to prevent glycerin from leaving the shell.

Banner Pharmcal and R.P. Scherer are the two most prolific custom manufacturers of soft shell capsules in the U.S.

Some solids are stabilized in oil solutions or in solutions of polyethylene glycol (e.g., several prostaglandins), so that soft shell capsules are a good dosage form for such products. Furthermore, dyes can provide light screening in the correct wavelength range and protect light sensitive entities.

Titanium dioxide is used as the most common opacifier. (Iron oxides are also used.) The titanium dioxide is opaque and (if considered completely non-transparent) will lengthen the pathway of light passing through the capsule (since the light will bounce from surface to surface of the opacifier), so that more light will be absorbed hence the product will be better protected from light.

13.6 SOFT SHELL CAPSULES

The preformulator, therefore, should determine:

1. the solubility of the compound in PEG or triglyceride;
2. the partition coefficient of the drug between capsule shell and fill-solution or suspension;
3. the stability of the compound in the fill-solution;
4. the stability of the compound in the shell;
5. since moisture will seep from the capsule shell into PEG fills, solubility and stability of the drug in PEG containing e.g. 5% of water should also be determined;
6. dissolution aspects. Although this cannot be adequately assessed at the early preformulation stage, it should be

determined whether aldehydes or ketones are present in the formulation, and whether these can interact with (and cross-link) the gelatin.

Such studies are best done by placing the drug in a gelatin solution, sampling it from time to time, drying it, and performing an IR spectral analysis on it. The H_2O stretching band is of no use, but the appearance of double bonds (caused by the condensation reaction between CHO or CO and NH_2 or NH) may reveal whether such reactions are possible. In that case "pellicules" or "pellicles" may form in dissolution testing of aged samples.

It is recommended that, for new NDA's, a type III dissolution apparatus be used, and, in case of pelliculation, that an in vivo test be carried out on pelliculated capsules, to ascertain that these are still bioavailable.

13.7 TABLETS

The manner in which tablets are made, on a single punch machine, is that the powder flows from the hopper (the shoe) into the die cavity, the bottom of which consists of the lower (movable) punch. The shoe then moves back, and the upper punch comes down and compresses the powder into a tablet. Both punches then move up and the tablet is ejected. The shoe then moves in over the punch again, knocking the tablet away, the lower punch lowers into its lowest position, and the cycle repeats. If the lowest position of the lower punch is raised, then less powder flows into the die cavity, hence the "fill weight" is decreased. In this manner it is possible to control the weight of a tablet. The lower the lowest position of the upper punch, the harder the tablet will become. The hardness is determined with a hardness tester, and the dimensions of the tablet are determined with a caliper or micrometer. There is a limit to how hard a tablet can be compressed, since if the position is too low, then the machine will freeze up.

In large scale, tablets are made on rotary machines.

Not all powders may be compressed in a satisfactory manner. Some powders will not flow, and hence will not fill up the die cavity. This may be demonstrated by flow of powders through a hopper. The granular powder flows well (fast), the fine powder does not flow well. It is noted that the powder, once it flows out of the hopper, forms a cone (a heap) on the supporting table. The angle of this cone is called the repose angle, α, and is characteristic of the powder. To make powders flow it is often necessary to make their particle sizes larger. This is done by wet granulation which will be demonstrated below. Some powders, however, can be compressed without further operations, and they are called direct compression excipients. Some of these are shown in the lab:

Dicalcium phosphate (Ditab)
Tricalcium phosphate (Tritab)
Spray-dried lactose
Microcrystalline cellulose (Avicel)

If a drug is potent, and only small amounts are added to a tablet, then it may be admixed with one of the above, and tableted directly. If not, then if it does not flow well or compress well, it is wet granulated. This is demonstrated with a so-called lactose-starch granulation. The formula for this is shown below:

Drug (FDC Yellow Dye #5)	6 g
Cornstarch USP added dry	180 g
Lactose USP	1515 g
Cornstarch USP (for paste)	45 g

The cornstarch is suspended in 100 mL of cold water, and added to 350 mL of boiling water. This forms a paste. The dye is blended with 115 g of lactose in a bottle (preblending). The remaining powders are added to the bowl of a planetary mixer, and mixing carried out for a couple of minutes. The drug (dye) preblend is then added, and mixing carried out for 5-10 minutes. If some powder is taken out and smeared out with a spatula, it may be seen that mixing is not complete (or when it is complete). The starch paste is now added. It is noted how it spreads out in a few moments, and the mixture goes from lumpy and non-uniform to quite uniform. This is now passed through a #6 hand screen, and the granules now, although wet, are quite uniform. These are then dried in tray drier. The dried granulation is too coarse, and is hence milled, mostly in a hammer mill.

If the powder were not milled, then it would be difficult to get an even edge in the compression step at the point where the shoe moves away, and this would affect the accuracy and precision of the fill.

To ascertain that the powder is fine enough it is subjected to a sieve analysis (as described below).

13.8 MECHANICAL TESTING OF SOLID DOSAGE FORMS

Hardness of Tablets

The hardness of a tablet is, if the drug is potent and most of the tablet is excipient, a function of the excipient, and as such the mechanical properties of the drug substance are of less impact on the tablet unit. If, however, there is more than 10-15% of drug in a tablet, then the mechanical properties of the drug substance become most important.

Yield Value or Elastic Limit ϕ

The hardness of a tablet is a function of how much pressure has been exerted in making it. The pressure needed to make the particles bond is the elastic stress limit or the yield value of the powder, and the upper limit of the pressure that can be applied is the so-called capping limit.

The preformulator can, with even small quantities of powder, determine the value of ø, in the following fashion: If small compacts are made at say 5 compression forces (2-10 ton force), then the thickness can be measured for each of these. They could be as shown in Figure 13.4.

The tablet in the case in Figure 13.4 has a cross-section that (for simplicity) is 1 cm^2, so that the volume of each tablet in cm^3 is 1/10 of the thickness (which is in mm).

The volume corresponding to a "fully" compressed compact can be calculated from the true density, and if it were 0.8 cm^3 then each thickness can be converted to porosity, ε, by the formula:

$$\varepsilon = 1 - (V_t/V)$$

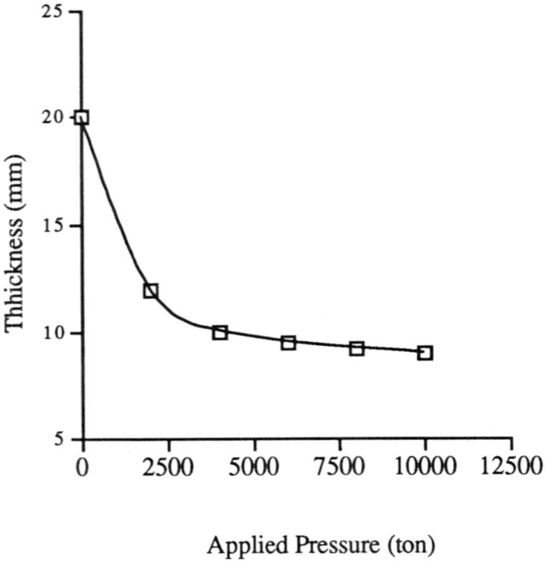

FIGURE 13.4 Thickness of a tablet as a function of applied pressure.

This is done in Figure 13.5 and the curve has the equation (Athy, 1930):

$$-\ln[\varepsilon] = 1.238 - 0.0001P \qquad (13.1)$$

The slope of the line is ø/3 (Heckel, 1961), so that the yield value (in stress units) in this case (since the cross-sectional area is 1 cm^2) is 0.0003 $tons/cm^2$.

The intercept is the porosity of the powder bed as it is positioned in the tablet die.

At high pressures the tablet will start to cap, and the hardness will go down again. This is a common type of defect. The reason for it occurring is that when the upper punch starts

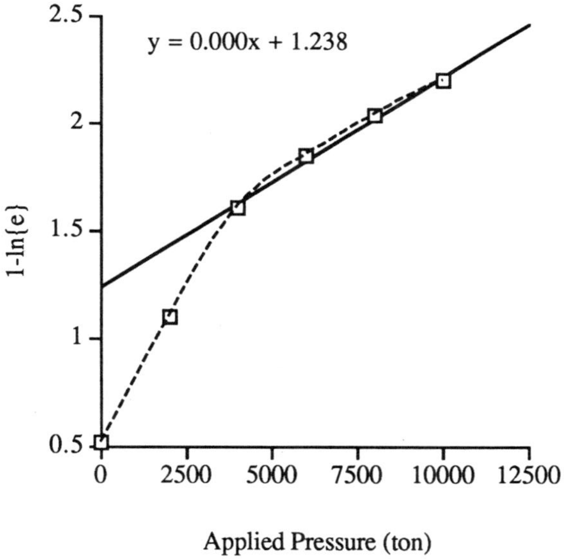

FIGURE 13.5 Data from Figure 13.4 plotted according to the Heckel equation. Note that the truncation in the equation is too severe, and that the slope is 0.0001.

"going up" again, the pressure is released, and the tablet can expand in only one direction (uniaxial expansion). This causes stress on the tablet, and although the applied pressure makes the tablet stronger, at a given point the extra strength acquired by the additional tableting pressure is offset by the additional stress in the expansion.

Figure 13.4 shows a typical hardness/pressure graph.

13.9 DISINTEGRATION TEST

For common tablets to be effective they must disintegrate (in fact they must usually disintegrate before they dissolve), and they are therefore subjected to a disintegration test routinely.

Tablets (Figure 13.1) are placed in tubes that have, as a bottom, a 10 mesh screen. They are placed in water and oscillated (up and down), and the time it takes for the tablet to fall apart is denoted the disintegration time.

13.10 TABLETS - DISSOLUTION

When a tablet is exposed to dissolution, it must first (a) wet, then (b) disintegrate. The particles that are formed are not prime particles, and water must then penetrate the granules before drug can diffuse out. The length of time, t_i, required for this is the lag time. The remainder of the curve becomes a diffusion type curve, i.e.

$$\ln[M/Mo] = - k\{t-t_i\} \tag{13.2}$$

k and t_i can be found from slope and intercept when the fraction not dissolved [M/Mo] is plotted versus time.

Examplewise the data in Table 13.1 are obtained and the data are plotted in Figure 13.6.

TABLE 13.1 Dissolution Data

Time	Fraction Released	ln[M/M_0]
0.000	0.000	
5.000	0.000	0.000
10.000	0.330	-0.400
15.000	0.551	-0.800
20.000	0.700	-1.200
25.000	0.800	-1.600
30.000	0.875	-2.000
35.000	0.910	-2.400

13.11 PHYSICAL COMPATIBILITY

Historically, preformulation had its birth in compatibility testing. The most problematic situation of this kind is when a drug substance "melts" in the presence of an excipient.

Suppose a compound has a melting point of 135°C and a heat of melting of 500 cal/mole, then a program as shown in

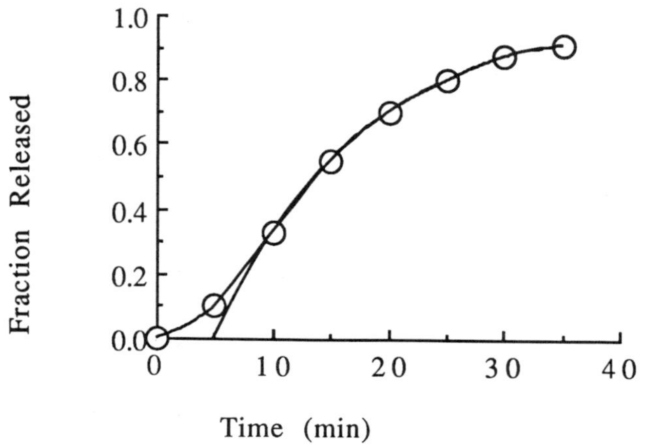

FIGURE 13.6 Data from Table 13.1.

TABLE 13.2 Program for Generating a Eutectic Program

```
100 INPUT "Heat of Fusion=";Q1
110 INPUT "Pure Melting Point=";T1
120 FOR X1 = 0 TO 1 STEP .1
130 Y1 = Q1/1.99
140 Y2 = LOG(1-X1)
150 Y3 = -Y2/Y1
160 Y4 = 1/T1
170 Y5 = Y4 + Y3
180 Y6 = 1/Y5
190 PRINT X1,Y6
2000 NEXT X1
```

Table 13.2 can be run, and the data in Table 13.3 will be generated. If compound B has a melting point of 110°C and a heat of fusion of 600 cal/mole, then the data in Table 13.4 will be generated. It should be noted that what in the program is denoted x is the mole fraction of B, and for the right hand side of the curve, this becomes (1-x). The data from Tables 13.2 and 13.3 are plotted in Figure 13.7.

TABLE 13.3 Eutectic Data for Compound A; ΔH = 500 cal/mole

Mole Fraction, x	Melting Point,°C	1 - x
0	135	0
0.1	127.6	0.9
0.2	120.5	0.8
0.3	113.3	0.7
0.4	105,8	0.6
0.5	98.4	0.5
0.6	90.5	0.4
0.7	82.0	0.3
0.8	72.4	0.2
0.9	60.3	0.1

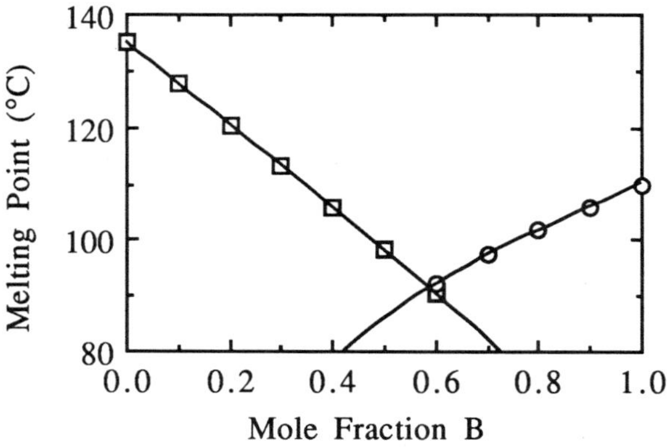

FIGURE 13.7 Data from Table 13.2 and from Table 13.3 showing the determination of a eutectic point (90°C) at the eutectic composition, 0.6 mole fraction B.

13.12 SOME ROUTINE TESTS

Several tests are usually carried out on powders in phase I and II. These are important, because they document the state of matter of the drug substance used in these stages.

13.13 WEIGHT AND CONTENT UNIFORMITY

Ten tablets are weighted on an analytical or a top-loading balance. There is a USP test that must be met, determining how much variation there can be between tablets.

The dissolution of a drug, as will be discussed, is, among other factors, a function of the average particle size of the drug. Tablets are subjected to assay as well, and content uniformity carried out according to USP.

TABLE 13.4 Eutectic Data For Compound B; ΔH = 600 cal/mole

x	(1-x)	Melting Point
0.0	1.0	110
0.1	0.9	105.9
0.2	0.8	101.7
0.3	0.7	97.3
0.4	0.6	92.7
0.5	0.5	87.8
0.6	0.4	82.4

Content uniformity is primarily a problem associated with (a) cohesion and (b) particle size distribution. In certain instances it is possible to attain very good and robust content uniformity with a micronized drug if spray-dried lactose is used as a diluent. In such cases the micronized drug should have a surface area of above 3 m^2/g. This assures an ordered mixture that has no tendency to segregate.

13.14 APPARENT DENSITY OF MIXTURES

At a point in development where more drug substance is available than at the initial preformulation stage, the preformulator may be concerned with the actual filling of the drug substance into the solid dosage form. Hard shell capsules come in different sizes and the amount of drug-excipient mixture that may be filled into a capsule depends on the apparent density, ρ^* , of the mixture. This in turn is a function of the true densities, ρ_o, and the apparent densities, ρ', of the components of the powder blend.

TABLE 13.4 Apparent Densities of a Binary Mixture

Fraction B	Apparent Volume, mL/g	Apparent Volume, g/mL
0	1.25	0.800
0.1	1.15	0.870
0.2	1.05	0.952
0.3	1.00	1.000
0.4	1.00	1.000
0.5	1.09	0.917
0.6	1.18	0.847
0.7	1.27	0.787
0.8	1.36	0.735
0.9	1.45	0.690
1.0	1.54	0.649

Apparent densities of powder blends may be obtained easily by pouring the mixture into a graduated cylinder to a given mark (e.g., 10 mL) and obtaining the solids weight by weighing. Mixtures of different compositions may be determined in this fashion, to obtain apparent density as a function of composition.

If a capsule size has a volume V cm^3, then the amount that may be filled into it would be $V\rho^*$. It can be shown that the apparent densities of a powder mixture of two powders A and B depend on the fraction of B as shown in Figure 13.8 and that this is linearized by taking inverses (Figure 13.9).

The theoretical treatment of this is based on the assumption that the particles are spherical. Deviations in slopes of the lines are a function of the shape factor of the particles.

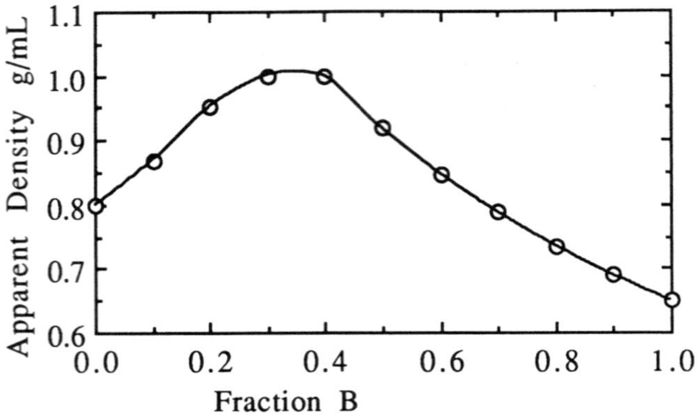

FIGURE 13.8 Data from Table 13.4.

13.15 COHESION - REPOSE ANGLES

The correct manner of testing for cohesion is by way of a shear/stress locus as described in the previous chapter. However, this requires a large amount of material, in general. At a point in development where enough material is available it is worthwhile carrying out a repose angle both on the raw material and on the finished blend.

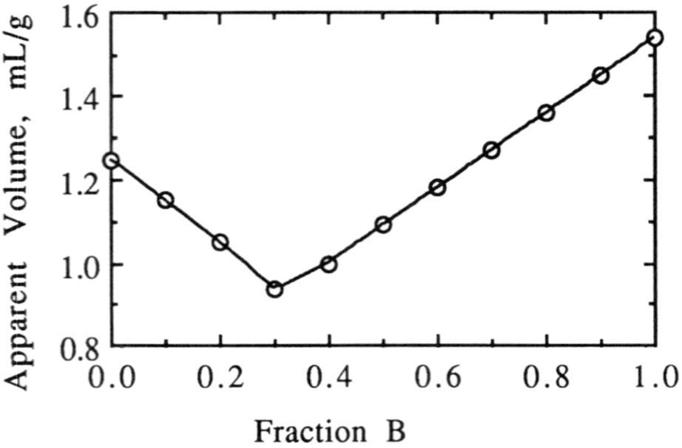

FIGURE 13.9 Data from Table 13.4.

The repose angle of the drug should be correlated with (a) the fractal dimension and (b) the particle distribution. This will provide a library of figures that can be tied in with product performance.

13.16 SIEVE ANALYSIS

Particle size distribution assessment has been discussed in a previous chapter. At this point it should be noted that powder for processing (tableting, encapsulation) should always be subjected to sieve analysis. In this test, as mentioned, powder is placed in the top screen of a nest of sieves. Each sieve has been weighed (tared) prior to the sieve test. The size of the opening is in "U.S. mesh" (ø) and the openings are finer the further down in the nest the screen is. The opening between the wires is about 2.54/(2.ø) cm. 100 g of powder are placed on the top screen and then vibrated for 5 minutes, and the individual sieves taken apart and weighed.

The weights are then tabulated, and it will be shown later in the course how the average particle diameter is calculated.

Aside from this, powder flow should be recorded, and this also tied in with product performance. Both powder flow and tablet weight variation are a function of particle size distributions.

13.17 EFFECTS OF POLYMORPHISM

The physico-chemical aspects of polymorphism have been discussed in Chapter 7. Making and identifying polymorphic forms fall within the realm of the preformulator. But aside from that, this aspect of a drug substance also affects qualities such as dissolution, apparent density, particle shape and flow.

Since the apparent solubilities of metastable polymorphs and amorphates are higher that those of the more stable forms, this aspect becomes of importance in dissolution testing and in hygroscopicity.

The metastable and amorphous forms are also more reactive, and conclusions regarding chemical stability in the solid state may be affected by this.

It can be shown that a dosage form of a moisture-sensitive compound will decompose in the presence of moisture by the following equation (Carstensen, 1983):

$$dM/dt = -k_1 S V \qquad (13.3)$$

where M is the amount of intact drug remaining at time t, S is the solubility of the compound in water, k_1 is the first-order degradation rate constant in aqueous solution, and V is the amount of water in the system.

Hence, a metastable polymorph or an amorphate will decompose more rapidly in the presence of water than the more stable form.

Moisture isotherms of the metastable forms exhibit a lower critical moisture level, and they may convert to the more stable form during the hygroscopicity experiment.

13.18 PREFORMULATION TASKS

Most of the text above is of an informative nature, and the preformulator would not be expected to carry out the tests, but should know the general principles, since his thinking should be geared towards aiding the formulator.

There are, however, some tasks of this nature that fall under the auspices of preformulation. The chemical stability of the dosage form should be tested in the presence of moisture. It should be ascertained that the polymorphic form of the drug substance does not change after 3 months' storage at 40°C and 75% RH. When sufficient material is at hand, the apparent density of the powder should be tested in different ratios with the most predominant excipient. Such results should be recorded for several batches and the results analyzed in light of the fractal dimension and the shape factor of the particles. Eutectic diagrams should be carried out by differential scanning calorimetry for mixtures of the drug substance with the most predominant excipients. Heckel plots should be carried out (e.g., on a hydraulic press with a pressure gauge) to determine the yield strength of the drug substance. A shear cell analysis should be carried out when sufficient sample is available.

REFERENCES

Athy, L.F., (1930), *Bull, Am. Assoc. Petrol. Geologists,* **14:**1

Carstensen, J.T., (1983), *Pharmaceutics of Solids and Solid Dosage Forms*, Technomic Publishing Co., Lancaster, Pa., p. 107.

Heckel, R.W, (1961), *Trans, Metal Soc. of A.I.M.E.,* **221:**671

CHAPTER 14

Compatibility Testing

When a program is started to investigate a new drug substance, the preformulator is, as mentioned earlier, usually presented with a very small amount of drug substance.

Since the drug substance has to be formulated into a dosage form for use in clinical trials, and if the clinical results so warrant, eventually into a marketed product, it is important to know which raw materials may be used in combination with the drug.

There are many examples of chemical drug-excipient interactions that lead to poor stability of the drug substance in dosage forms using that particular excipient, and such information should always be extracted at the earliest point in the development program.

14.1 COMPATIBILITY TEST FOR SOLID DOSAGE FORMS

One manner in which this may be done is to make a small mix of excipients and the drug substance. This is placed in a vial, stoppered with a rubber stopper which is then dipped in molten carnauba wax. This latter step renders the system hermetically sealed, although another method is placing it in a vial and then sealing the vial. The important point is to keep the system closed to the outside. The wax hardens on cooling and forms a moisture barrier which is effective up to a temperature of 70°C.

Common excipients which may be used in this type of test have been reported by Carstensen et al. (1964) and by Perrier and Kesselring (1983).

The above test is carried out with drug substance and excipients as is, but a similar set of samples may be set up with moisture, e.g., 5% moisture added. Two weeks storage at 55°C is a condition that may be employed. An exception is stearic acid and dicalcium phosphate, where 37°C is more appropriate (Toy, 1980) because of, in the case of stearic acid, the low melting point, and in the case of dicalcium phosphate, its tendency to dehydrate at temperatures above 40°C. After storage the sample is observed physically for the following: (1) liquefaction, (2) caking, (3) odor or gas formation and (4) discoloration. It may then be assayed by HPLC or thin-layer chromatography (TLC).

Samples of the drug substance itself are always set up as a control. Aside from allowing assessment of whether a new spot on a TLC (or peak in a HPLC) is due to the drug or the excipient, it is by now an FDA requirement of sorts (Schultz, 1984).

Furthermore, it allows the preformulator to disregard spots that might have originated in the batch of raw material itself. The important point is whether the drug (or excipient) broke down during the test, not whether it was somewhat decomposed or contained impurities at the onset.

The method, hence, allows selection of excipients that are compatible with the drug substance, and the criterion used is that no new spots on the TLC chromatogram have developed after the accelerated storage.

At times, liquefaction occurs due to eutectic formation Examples are drug-caffeine combinations which often exhibit this phenomenon. In this case there may not be a chemical incompatibility, but the physical change may make the physical characteristics of, for example, a tablet undesirable, and the combination is therefore considered an incompatibility.

On the other hand there is a series of known chemical incompatibilities known to most preformulators and formulators. Sugars and primary amines, for instance, are usually now compatible, and magnesium stearate is incompatible with many systems, aspirin being one classical example of this.

14.2 HYDRATES

Some drug substances and excipients exist as hydrates. Hydrates have been discussed in a previous chapter, but some preformulation considerations are covered at this point.

Some compounds may exist as a series of hydrates, e.g., Na_2HPO_4 which may exist as an anhydrate and as a dihydrate, a hepta-hydrate and a dodecahydrate. It is assumed in the following that the drug exists as an x-hydrate:

$$A.xH_2O \rightarrow A + xH_2O \qquad (14.1)$$

The equilibrium constant, K, for this (Maron and Prutton, 1965) is given by:

$$K = a_A a_{H2O}{}^x / a_{xaq} = a_{H2O}{}^x = P_{H2O}{}^x \qquad (14.2)$$

The temperature effect on this is given by:

$$\ln P = (-\Delta H/xR)(1/T) + Q \qquad (14.3)$$

The vapor pressure for $Na_2HPO_4.7H_2O/Na_2HPO_4.12H_2O$ is given in Table 14.1. These data are shown in Clausius-Clapeyron fashion in Figure 14.1. This has an impact on how high temperature stability studies are carried out, since a hydrate will lose enough moisture in a closed container to achieve the required vapor pressure in the head space. *Often the anhydrate formed by dehydration of a hydrate will be amorphous and show decreased stability over that of a crystalline anhydrate.* Typical examples of this are amoxicillin trihydrate and ampicillin trihydrate. In these cases, dehydration of the trihydrate gives rise to a chemically unstable, amorphous anhydrate. Furthermore, storage should be in glass, rather than in polyethylene (or other polymer), since moisture loss will cause aberrant stability behavior. It is, therefore, important to establish whether a drug substance exists as a hydrate or simply contains a certain amount of moisture. There are several practical ways of answering this question.

TABLE 14.1 The Na_2HPO_4 Salt Pair System

Temperature °C	Water Vapor Pressure (Torr)
0	2.68
15	8.97
20	13.0
25	19.2
30	27.1

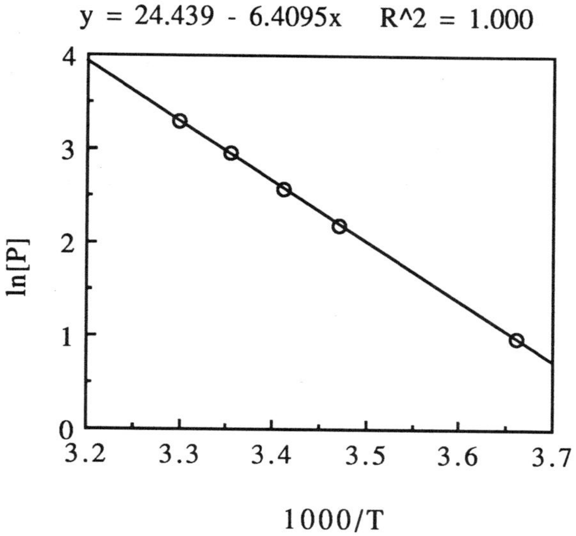

FIGURE 14.1 Data from Table 14.1.

(a) If the compound with a given moisture content (b_1) is placed in a closed container, and a RH-measuring device is placed in the head space, then, upon equilibrium, a given RH (a_1) is registered.

This is now repeated with several moisture contents, and if identical RH-values are obtained for a series of moisture contents, then a salt-pair exists. A diagram of a situation where a monohydrate and a dihydrate exist is shown in Figure 14.2.

The next point then is to establish the stoichiometry (e.g. that the compound is probably a dihydrate). This is then confirmed by preparing samples with more than 2 moles per mole of dry solid, and registering a higher RH.

It should be pointed out that equilibrium takes place slowly, and that several months may be necessary for conclusions to be made.

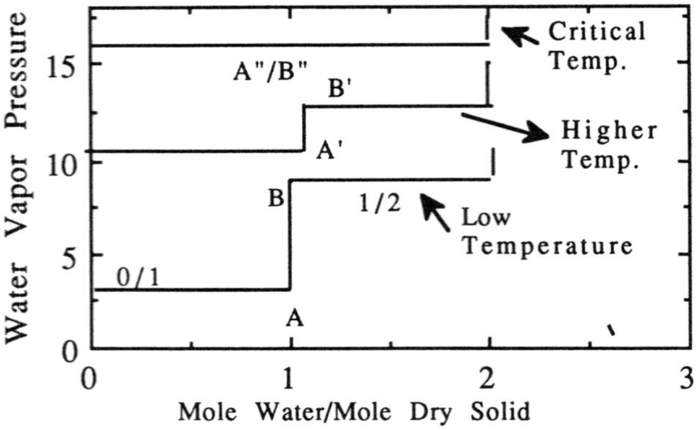

FIGURE 14.2 Vapor pressure diagram for a compound forming two hydrates, showing the anhydrous/monohydrate and the monohydrate/dihydrate salt pairs. The distance AB decreases with increasing temperature, and at the critical temperature they coincide. Position of isobars is arbitrary in this presentation (see e.g., Figures 14.3 and 14.4).

(b) Samples may be placed in desiccators at different RH-values, and by weight it may be established at which RH moisture is given off, and at which it is taken up.

The same method may be carried out in VCI-type instruments, where the instrument is kept at constant RH for a fairly long time (e.g., days) to allow for equilibration.

Again, uncertainties about whether equilibrium has been reached exist.

(c) Of importance is establishment of the RH range at which a hydrate is stable at different temperatures (Carstensen, 1986). The decomposition temperature may be established by DSC. This will establish the enthalpy of the dehydration reaction of the salt.

The hydrate will have a lower vapor pressure than its saturated solution (P_{sat}). The decomposition temperature of a salt is the temperature at which the saturated solution has the same vapor pressure as the hydrate.

A delicate situation is that of calcium phosphate dihydrate which has a fairly low decomposition temperature. It may show up as being incompatible in many products at higher temperatures (because of water loss), but the product may be quite stable at room temperature (20-25°C, per USP).

The ICH guidelines prescribe a 40°C, 75% RH test, and dicalcium phosphate combination that at times unfairly affects the use of dicalcium phosphate dihydrate as an excipient. In cases where it fails it is recommended that the ICH fall back position of 30°C and 60% RH be used.

The effect of temperature on salt pairs is illustrated in Figure 14.1 where the compound is illustrated as forming two hydrates, a mono and a dihydrate. It is noted that the vapor pressure is constant when there is presence of both hydrates in the salt pair.

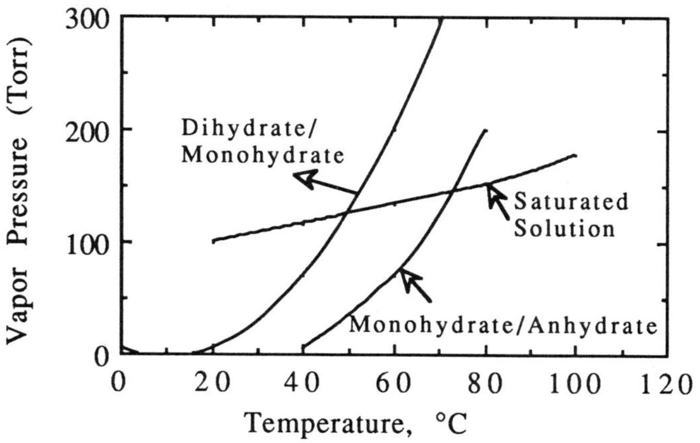

FIGURE 14.3 Vapor pressures of salt-pairs as a function of temperature, where there is no transition.

The decomposition temperature of a hydrate is the temperature where the vapor pressure of the saturated solution becomes equal to that of the hydrate. This is exemplified in Figure 14.3. It should be noted that in the case where more than one hydrate exists, there can be transition temperatures (Brønsted, 1943) as shown in Figure 14.4. Here there are two salt pairs; at temperatures below A (72°C) they are monohydrate/anhydrate and trihydrate/monohydrate. Above 72°C the curves would suggest that the monohydate would have a higher vapor pressure than the trihydrate, which is thermodynamically impossible. 72°C then becomes the decomposition temperature for the monohydrate, and above A (72°C) only the salt pair (trihydrate/anhydrate) exists. The trihydrate then becomes unstable when its vapor pressure equals that of a saturated solution of trihydrate (not shown in Figure 14.4). It should be noted that the salt pair trihydrate/anhydrate is not stable below A (72°C), only above.

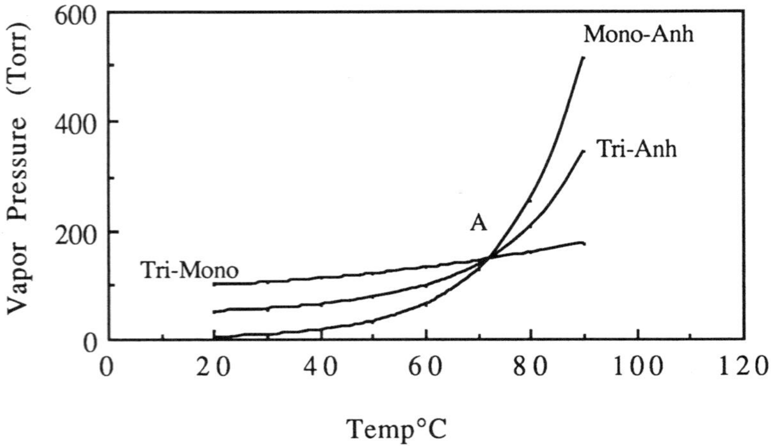

FIGURE 14.4 Case of a compound with two salt pairs where there is a transition point.

14.3 DISSOLUTION COMPATIBILITY

Dissolution is of importance from a biopharmaceutical point of view. In general, as a rule of thumb, if the solubility is greater than 1 mg min^{-1} cm^{-2}, the drug is not prone to give dissolution rate limited absorption problems. On the other hand, if the value is less than 0.1, such problems can definitely be anticipated, and compounds with values of kA/V of from 0.1 to

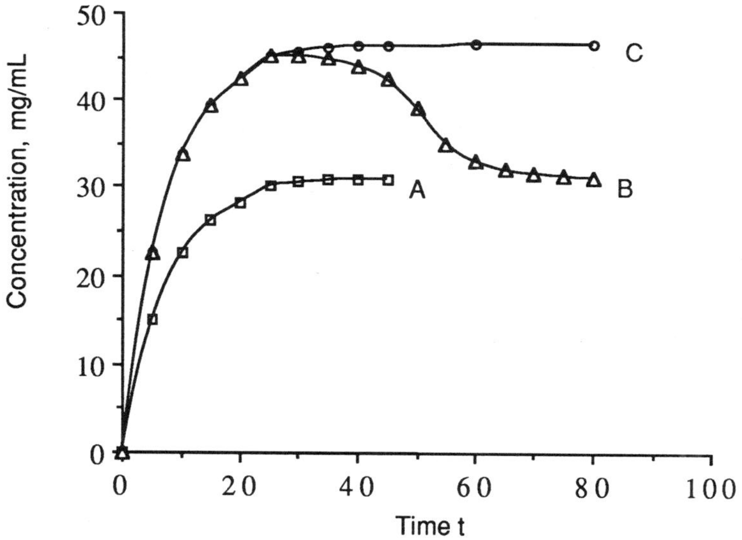

FIGURE 14.5 Dissolution profiles obtained from the solubility determination of two polymorphic forms of the same drug substance. A is the stable form with solubility 31 mg/mL. B is the profile of the metastable form with solubility 46 mg/mL. This solubility, C, (circles) is not achieved in many instances, and precipitation of the stable form occurs at a point beyond the solubility of A, and the trace becomes B.

1 mg min^{-1} cm^{-2} are in a gray area. For compound selectivity it is frequently useful to express dissolution findings in terms of k (i.e., in cm/sec).

For a small amount of powder, dissolution of the particulate material can often be assessed (and compared with that of other compounds) by placing the powder in a calorimeter (Iba et al., 1991) and measuring the heat evolved as a function of time. The surface area must be assessed microscopically (e.g. by image analyzer), and the data must be plotted by a cube root equation (Hixson and Crowell, 1931):

$$1 - [M/M_o]^{1/3} = -(2kS/\rho r)t \qquad (14.4)$$

where M is mass not dissolved, M_o the initial amount subjected to dissolution, ρ is true density, S is solubility, and r is the mean "radius" of the particle. The method is simply comparative, not absolute, due to the hydrodynamics being different in the calorimeter than it would be in a dissolution apparatus.

Dissolution testing (in effect solubility testing) is an early indicator of the possibility of polymorphs. It has been established in an earlier chapter that the metastable polymorph has a higher its solubility. When the metastable polymorph dissolves, it will first approach this higher solubility, but then at one point the more stable form will nucleate, and precipitation of this form will start, so that the solubility eventually levels off at that at a lower level. Figure 14.5 is an example of this. In order to obtain the solubility of the metastable polymorph in such a case, the Nogami method (Nogami, 1966) is used, as demonstrated for the data in Figure 14.5. The data for the initial portion of curve C are shown in Table 14.2.

It is noted in Figure 14.5 that equal time intervals are used (Δt, $2\Delta t$, $3\Delta t$ etc.). If C at Δt is plotted versus C at $2\Delta t$, and C at $2\Delta t$ is plotted versus C at $3\Delta t$, then a plot such as shown in Figure 14.6 results. The solubility of the metastable polymer is the intersection of the line with the line y=x.

TABLE 14.2 Dissolution Data from Fig. 14.5

Time	C at Time t	C at t+Delta
0	0	22.5
5	22.5	33
10	33	39
15	39	42
20	42	

14.4 VAPOR PRESSURE

In general, vapor pressures are not all that important in preformulation, but it should always be kept in mind that a substance may have sufficiently low vapor pressure to (a) become lost to a sufficient extent to cause apparent stability problems, and content uniformity problems and (b) exhibit a potential for interaction with other compounds and adsorption onto or sorption into package components (Pikal and Lukes, 1976).

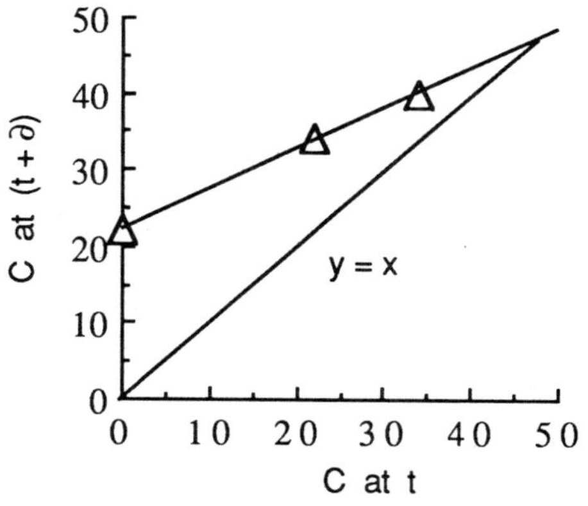

FIGURE 14.6 The Nogami method applied to the upper curve in Figure 14.5. The least squares fit for the upper line is y = 27.081 + 0.53139x so that y = x when x = y = 22.081/(1-0.53139) = 47.3.

Most drug substances are, substantially, not volatile. As an initial screen, it can be determined whether the drug is sufficiently volatile to cause concern by placing a tared amount of it in a vacuum desiccator and weighing it daily for a while. It is better to have a high-vacuum system for this, and the use of a vacuum electrobalance is best for this purpose. The ratio of the slopes is A_bP_b/A_aP_a so if the specific surface areas is known for each, then (since P_b of the reference is known), P_a can be calculated.

A good estimate of the vapor pressure can be obtained (Carstensen and Kothari, 1981) by using a pierced thermal analysis cell, placing it on a vacuum electrobalance, and monitoring the weight loss rate. Using a substance with known vapor pressure can then be used for calibration, the loss rates being proportional to the vapor pressures.

By using constant temperature TGA, graphs such as shown in Figure 14.7 will result. The weight rate (which should be established as due to evaporation of the compound) is given by:

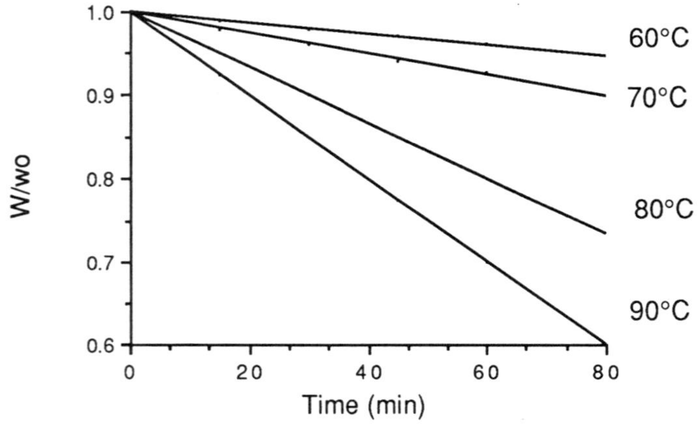

FIGURE 14.7 Weight loss curves from constant temperature TGA.

$$dW_a/dt = -kA'P_a \quad (14.5)$$

where W is weight, t is time, k is a heat/mass transfer coefficient, A' is surface area, and P_a is the vapor pressure of the compound at the given temperature.

If the specific surface area is

$$A_a = A'/W_{ao} \quad (14.6)$$

where W_{ao} is the original weight of the sample, so that this, inserted in Eq. 14.5 gives:

$$W_a/W_{ao} = 1 - kA_aP_at \quad (14.7)$$

An example of such a plot is shown in Figure 14.7.

If a compound with known vapor pressure (e.g. benzoic acid, subscript *b*) is subjected to the same conditions, then it will exhibit a weight loss curve given by a similar equation:

$$W_b/W_{bo} = 1 - kA_bP_bt \quad (14.8)$$

The heat of sublimation, ΔH, can be obtained by plotting the negative of the slopes by a Clausius-Clapeyron equation. In the cited case the negative of the slope of this plot is $\Delta H/R = 8.7$ kCal/mole, so that the heat of sublimation is $\Delta H = 17.4$ kCal/mole. For this, no reference is necessary; all that is necessary is to know and adjust for the weights of the samples studied.

14.5 AQUEOUS SOLUTION COMPATIBILITY

In general, such studies are carried out by placing the drug in a solution of the additive. These can be (and usually are) a heavy metal (with or without chelating agents present) or an antioxidant (in either oxygen or nitrogen atmosphere). Usually, both flint and amber vials are used, and in many cases an autoclaved condition is included. This will answer questions about susceptibility to oxidation, to light exposure, and to heavy metals. These are important questions as far as injectable compatibilities are concerned. Exposure to various plugs is frequently included at this point so that early injectable preparations can be formulated.

For preparations for oral use, knowledge of the desired dosage form is important, but compatibility with ethanol, glycerin, sucrose, corn syrup, preservatives, and buffers is usually carried out. This type of study also gives an idea of the activation energy, E, of the predominant reaction in solution. Arrhenius plots for compounds in solution are usually quite precise.

14.6 NON-AQUEOUS LIQUIDS

With transdermal dosage forms being of great importance of late, it is advisable to test for compatibilities with "ointment" excipients and with polymers (e. g., ethylvinyl polymer, if that is the desired barrier). In the case of transdermals, the dosage form is either directly placed in a stirred liquid, or it is placed in a cell with an appropriate membrane (e.g., cadaver skin) to estimate the release characteristics of the drug from the ointment (Chien et al., 1983).

It should be noted here that if the overall flux is J, then

$$1/J = [1/(J_{ointment})] + [1/(J_{membrane})] \qquad (14.9)$$

where subscripts refer to the respective phase. $J_{membrane}$ can be obtained from curves such as shown in Figure 14.8. First the overall flux is obtained (with the membrane in place), giving the value of J. Then the release is obtained without the membrane in place, giving $J_{ointment}$, that is,

$$J = [1/A]\ dm_1/dt \tag{14.10}$$

and

$$J_{ointment} = (1/A)\ dm_2/dt \tag{14.11}$$

$J_{membrane}$ is then obtained as the reciprocal of the difference.

In vivo testing is usually carried out by applying the dosage form to hairless rats followed by subsequent sacrifice. Since the skin consists of a number of layers with differing hydrophilicity, the overall fate of the drug is of importance.

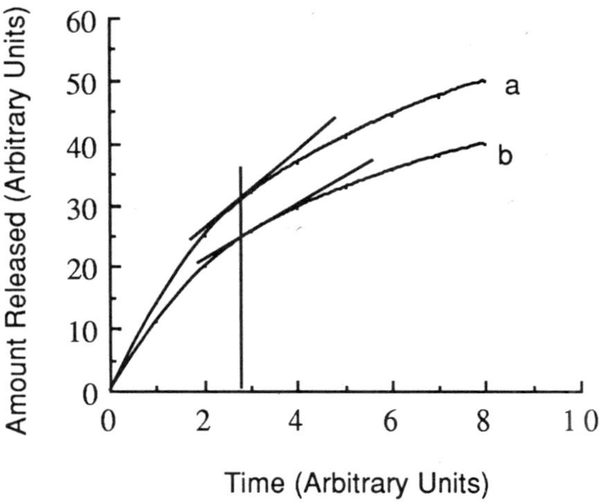

FIGURE 14.8 Slope determination of flux of ointment release and release from ointment + membrane.

14.7 PREFORMULATION TASKS

The individual subjects covered in this chapter should be tested, depending on the type of dosage form that is anticipated. For instance, liquid compatibilities are not necessary when only solid dosage forms are anticipated.

In carrying out solid two-component compatibilities, most preformulators consult the formulas usually used in the company to establish which excipients are usually used. Unless all of these show incompatibility, there is no pressing need to investigate others, at least not until a fair amount of drug substance is on hand.

For hydrates it is useful to store the hydrate(s) in desiccators at different relative humidities and at different temperatures to establish data of the type depicted in Figure 14.3.

It is useful to carry out a dissolution study over a long period of time (provided stability allows this). In this fashion it is often possible to establish whether the polymorph at hand is the more stable one.

It is worthwhile to estimate the volatility of a new drug substance, as described in Section 14.4.

REFERENCES

Brønsted, J.N., (1943), *Fysisk Kemi,* 2nd Edition, Munksgaard, Copenhagen, pp. 183-184.

Carstensen, J.T., Johnson, J.B., Valentine, W. and Vance, J., (1964), *J. Pharm. Sci.*, **53**:1050.

Carstensen, J.T. and Kothari, R., (1981), *J. Pharm. Sci.*, **70**:1095.

Carstensen, J.T., (1986), *Pharm. Tech.*, 98.

Carstensen, J.T., Franchini, M. and Ertel, K., (1992), *J. Pharm. Sci*, **81**:303.

Chien, Y.W., Keshary, P.R., Huang, Y.C. and Sarpotdar, P.P., (1983), *Drug Dev. Ind. Pharm.*, **72**:968.

Hixson, A. and Crowell, J., (1931), *Ind. Eng. Chem.*, **23**:923.

Iba, K., Arakawa, E., Morris,T. and Carstensen, J.T., (1991), *Drug Dev. Ind. Pharm.*, **17**:77.

Maron, S.H. and Prutton, C.F., (1965), *Principles of Physical Chemistry*, 4th Edition, MacMillan, Collier-Macmillan Ltd. London.

Nogami, H., Nagai, T. and Suzuki, A., (1966), *Chem. Pharm. Bull.*, **14**:329.

Perrier, P.R. and Kesselring, U.W., (1983), *J. Pharm. Sci.*, **72**:1072.

Pikal, M. and Lukes, A.L., (1976), *J. Pharm. Sci.*, **65**:1269.

Schultz, R.C., (1984), *Stability of Dosage Forms,* FDA-Industry Interface Meeting, Washington, D.C:, Oct. 7, 1983.

Stability Guidelines, *Congressional Record*, May 7, 1984.

Toy, A.D.F., (1980), Inorganic phosphorous chemistry, in *Comprehensive Inorganic Chemistry*, ed. Bailar, J.C., Jr., The Chemical Rubber Company, Cleveland, Ohio.

CHAPTER 15

Dissolution Test Design

Preformulators are often asked to design dissolution tests for products at an early stage in development. The idea is to have a dissolution test for the solid dosage form as soon as phase II or definitely when phase III clinical tests start.

Frequently tests will simply be run, and no actual thought is given to what the test signifies, other than it is a quality control tool, and that it eventually will become a quality control test post-NDA or post ANDA.

There have been suggestions of qualifying dissolution tests in the manner in which there is a "correlation" between in vitro dissolution and in vivo dissolution. The latter is often deduced by way of deconvolution.

The writing to follow will suggest a fairly simple method that will, in many cases, allow a meaningful in vitro to in vivo

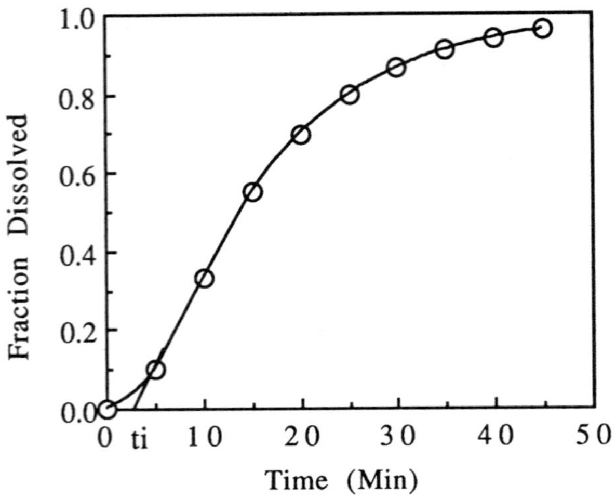

FIGURE 15.1 Typical dissolution pattern of a solid dosage form.

correlation. The FDA usually requires "gentle" tests which will distinguish between batches, and it is prudent to develop such a test and run it during clinical trials, but also to carry out a meaningful correlating test during development.

15.1 IN VITRO DISSOLUTION OF SOLID DOSAGE FORMS

Most tablets and capsules dissolve by a pattern as shown in Figure 15.1 (Carstensen, 1972, 1977).

The profile is characterized by a lag time, t_i, and a diffusional part. The lag time may be estimated by extrapolating the diffusional part down to the x-axis.

The equation for the curve is obtained by assuming that the concentration is zero up until t_i, and that it then becomes diffusional, i.e.

$$M/M_o = 1;\quad t < t_i \tag{15.1}$$

and

$$\ln[M/Mo] = -k(t-t_i); \quad t>t_i \tag{15.2}$$

M is here mass not dissolved, M_0 is the amount of drug at time zero, t is time, and t_i is the lag time.

Eq. 15.2 may also be written:

$$M = M_0 e^{-k(t-ti)} \tag{15.3}$$

or expressed as the amount dissolved, G:

$$G = G_0[1-e^{-k(t-ti)}] \tag{15.4}$$

where G_0 is the total amount of drug in the dosage form.

The data from Figure 15.1, treated according to Eq. 15.2, are shown in Figure 15.2.

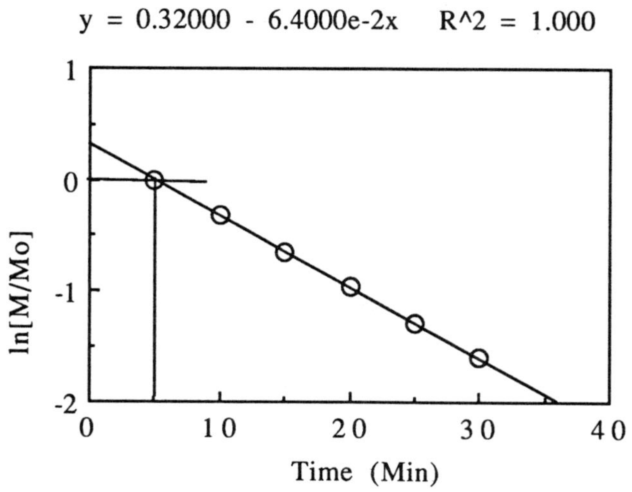

FIGURE 15.2 Data from Figure 15.1 treated by Eq. 15.2.

It is seen that the lag time is t_i = 5 minutes and that the dissolution constant (not to be confused with the intrinsic dissolution rate constant) is 0.064 min^{-1}.

15.2 BLOOD LEVEL EQUATIONS

Usually, if a solution is administered, the prevailing differential equations are:

$$dG(t)/dt = -k_aG(t) \tag{15.5}$$

and

$$dB(t)/dt = k_aG(t) - k_eB(t) \tag{15.6}$$

where

B(t) is the amount (not concentration) of drug in blood
G(t) is the amount (not concentration) of drug in the stomach (or GI tract)
k_a is the absorption rate constant
k_e is the elimination rate constant
t is time

There are assumptions made in this, for instance that the diffusional equations are based on amounts and not concentrations. This may, on the surface, seem minor, but it should be noted that the concentration may change not only because of absorption, but also due to dilution, i.e., the "volume" of liquid will change with time after administration, so that amounts are a more rational variable. There is also the assumption that k_a is a constant. These restrictions do not apply to the term B(t), because the volume of distribution may reasonably be considered constant, but it is emphasized, as well, that the model described is what is known as a one compartment model.

Nevertheless, the equations have been quite successful in the interpretation of blood level profiles obtained after administration of oral solution dosage forms. Eq. 15.5 has the direct solution:

$$G(t) = G_o(t)e^{-k_a t} \tag{15.7}$$

Inserting this in Eq. 15.6 then gives, after applying Laplace transforms:

$$\mathbf{L}\{dB(t)/dt\} = \mathbf{L}\{G(t)\} - k_e\mathbf{L}\{B(t)\} \tag{15.8}$$

where **L** is the Laplace transform of argument s. Noting that $B_o(t) = 0$, the expression:

$$\mathbf{L}\{dB(t)/dt\} = S\mathbf{L}\{B(t)\} \tag{15.9}$$

is used on the left-hand side of Eq. 15.8. The transform:

$$\mathbf{L}\{e^{-ft}\} = 1/(s+f) \tag{15.10}$$

is used on the right-hand side of the equation, so that Eq. 15.8 now becomes:

$$s\mathbf{L}\{B(t)\} = \{G_o/(1+k_a)\} + \{k_e\mathbf{L}\{B(t)\}\} \tag{15.11}$$

This rearranges to:

$$\mathbf{L}\{B(t)\} = G_o/(1+k_a)\,(s+k_e) = \{G_o/(k_a-k_e)\}\cdot[\{1/(s+k_e)\} - \{1/(s+k_a)\}] \tag{15.12}$$

The solution to this is then obtained by taking the inverse transforms:

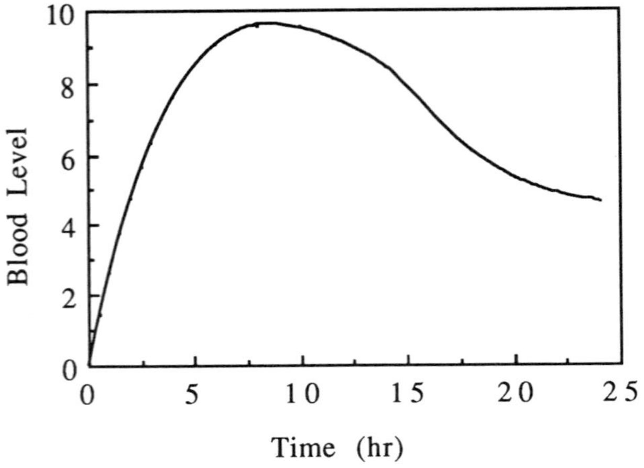

FIGURE 15.3 Trace of the function B(t) = exp(-0.1t) - exp(-0.12t) from 0 to 24 time units (hours, for instance).

$$B(t) = \{G_o/(k_a-k_e)\}\cdot[e^{-k_e t} - e^{-k_a t}] \qquad (15.13)$$

This type curve has been studied to a great deal in literature, and many examples of good fits exist.

A typical curve of this type is shown in Figure15.3.

15.3 IN VIVO DISSOLUTION

It is of interest to pharmaceutical scientists to attempt to imitate the in vivo behavior of a solid dosage form with in vitro methods, and one approach is the following: If a solid dosage form is administered, then it must first disintegrate, then dissolve, before it can be adsorbed. Hence the concentration (amount) profile in the stomach would be as expressed by Eq. 15.14:

$$dG(t)/dt = kA(t)M_s - k_aG(t) = Fe^{-qt} - k_aG(t) \qquad (15.14)$$

where

$$F = kA_oM_se^{qt_i} \qquad (15.15)$$

and A_o is initial surface area, A(t) is the surface area at time t, and M_s is the solubility amount. k_a is the absorption rate constant, and G(t) is the amount of material in solution in the GI-tract at time t. Sink conditions are assumed and it is assumed that the surface area decreases exponentially with time.

The Laplace transform of Eq. 15.14 is:

$$s\mathbf{L}(G(t)) - G_o = \{F/(s+q)\} - k_a\mathbf{L}\{G(t)\} \qquad (15.16)$$

or

$$\mathbf{L}(G(t)) = \{G_o/(s+k_a)\} + \{F/((s+q)(s+k_a)\} \qquad (15.17)$$

It is noted that G_o is not zero, because of the lag time. The differential equation for the amount of drug in the blood is the differential amount absorbed minus the amount eliminated (with rate constant k_b), i.e.:

$$dB(t)/dt = k_aG(a) - k_bB(t) \qquad (15.18)$$

Taking Laplace transforms and making use of Eq. 15.17 then gives:

$$s\mathbf{L}(B(t)) + B_o =$$
$$\{k_aG_o/(s+ka)\} + \{F/[(s+q)(s+k_a)]\} - k_b\mathbf{L}(B(t)) \qquad (15.19)$$

or:

$$L(B(t)) = B_o/(s+k_b) + \{k_aG_o/(s+k_a)(s+k_b)\} + \{F/((s+q)(s+k_a)(s+k_b)\} \quad (15.20)$$

The antiLaplacian to this is a triexponential:

$$B = A_1e^{-k_bt} - A_2e^{-k_at} - A_3e^{-qt} \quad (15.21)$$

where

$$A_1 = B_o + [k_aG_o/(k_a-k_b)] + [F/(k_a-k_b)(q-k_b)] - [F/(k_a-k_b)(q-k_a)] \quad (15.22)$$

$$A_2 = [k_aG_o/(k_a-k_b)] - [F/(k_a-k_b)(q-k_a)] \quad (15.23)$$

and

$$A_3 = [F/(k_a-k_b)(q-k_a)] \quad (15.24)$$

TABLE 15.1 Blood Level Data after Administration of a Tablet

Time (hours)	Blood Level
0	0
1	4.66
2	11.6
3	12.8
4	11.9
8	6.0
12	2.72
16	1.22
20	0.55
24	0.25

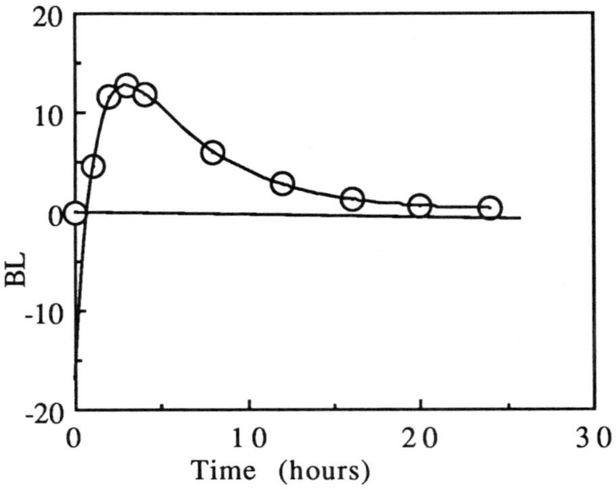

FIGURE 15.4 Data from Table 15.1. The drawn curve is the least squares fit of the data. Note that the initial value of the fit is negative, and that the point it intersects with the x-axis is the lag time.

The blood level curve may be fitted by non-linear least squares fitting to give the coefficients and the exponents. In this fitting the zero point should be excluded, and t_i will emerge as the point where the blood level curve crosses the time axis (x-axis). An example of blood level data is shown in Table 15.1 and Figure 15.4.

The best fit for the curve is:

$$B = 30e^{-0.2t} - 32e^{-0.8t} - 15e^{-1.0t} \tag{15.25}$$

It is noted that

$$B_o = 30-32-15 = -17 \tag{15.26}$$

and for t = 1 the calculated value is

$$B_1 = 4.7 \tag{15.27}$$

so

$$t_i = 17/(17+4.7) = 0.78 \text{ hours or } 47 \text{ minutes} \tag{15.28}$$

It is also noted that

$$q = 1.0 \text{ hr}^{-1} = 0.017 \text{ min}^{-1} \tag{15.29}$$

15.4 CALCULATING THE CORRELATING IN VITRO CURVE

If an in vitro curve shows a lag time of 5 minutes, then 5 minutes in vitro corresponds to 60 minutes in vivo. The ideal dissolution curve should have a dissolution constant such that

$$k_{\text{in vitro}} \times t_{i(\text{in vitro})} = q \times t_{i(\text{in vivo})} \tag{15.30}$$

Inserting the appropriate values now gives

$$k_{\text{in vitro}} = 0.017 \times 47/5 = 0.16 \tag{15.31}$$

The constructed ideal curve would then be:

$$\ln[M/Mo] = -0.16(t-5) \tag{15.32}$$

or

$$M/M_o = e^{-0.16(t-5)} \tag{15.33}$$

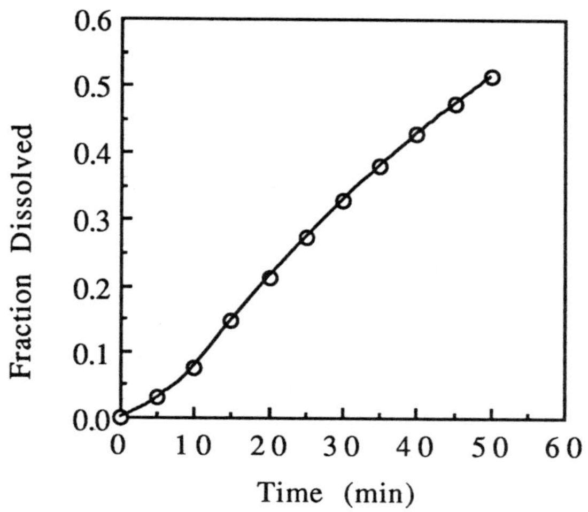

FIGURE 15.5 Trace of data adhering to Eq. 15.33.

A graph of this is shown in Figure 15.5. It is noted from the blood level curve in Figure 15.4 that the tablet does not dissolve rapidly in vivo, and that is the reason for the "slow" in vitro curve that corresponds to it.

In general, for immediate release dosage forms, "better" blood level curves are obtained, giving more rapid in vitro curves. The data are simply presented for the purpose of demonstrating the procedure.

15.5 ONE POINT METHODS AND DECONVOLUTION

There are other correlation methods. The literature is replete with procedures where correlations have been made by one point, e.g., the height of the absorption peak correlated with a Q_{30}-value. Such methods may have merit if they are tested in vivo, and faster and slower formulations are tested. In that manner one may correlate a curve or a value with a performance and state that a lower value or a higher value is unsatisfactory (Carstensen, 1980).

Conventional deconvolution may also be employed. In this method the blood level curve of a solution and a solid dosage form are compared, and the latter "subtracted" (by means of a deconvolution integral), so that the in vivo dissolution curve results. This can then be compared with the in vitro curve and correlation established (Saguwara et al., 1994).

SYMBOLS

$A(t)$ = surface area of solid at time t
A_o = initial surface area.
$G(t)$ = the amount (not concentration) of drug in the stomach (or GI tract) at time t
G_o = the value of G at time zero
k_a = the absorption rate constant ($k_a > k_b$)
k_b = the elimination rate constant
k_i = intrinsic dissolution rate constant
L = Laplace transform
$M(t)$ = mass of solid not dissolved at time t after the dissolution has started
s = argument of Laplace transform
t = time
t_i = lag time
q = exponential for area decay

REFERENCES

Carstensen, J.T., (1972), *Dissolution Technology*, Ed. Leeson, L.J. and Carstensen, J.T., IPT of the APhA, Washington, D.C., p. 192.

Carstensen, J.T., (1977), *Pharmaceutics of Solids and Solid Dosage Forms,* Wiley-Interscience, New York. p. 172

Carstensen, J.T., (1980), *Solid Pharmaceutics, Mechanical Properties and Rate Phenomena,* , Academic Press, N.Y., p.231

Sugawara, S., Imai, T., and Otagiri, M., (1994), *Pharm. Res.*, **11:**272

SUBJECT INDEX

E

F

G

H

I

K

L

M

N

O

P

R

S

T

V

W

X

Y

Z

AUTHOR INDEX

A

B

C

D

E

K

L

M

N

O

P

R

S